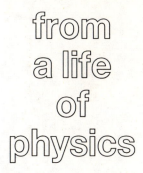

from
a life
of
physics

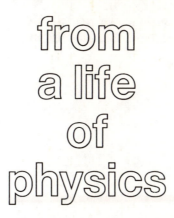

from
a life
of
physics

H.A. BETHE
P.A.M. DIRAC
W. HEISENBERG
E.P. WIGNER
O. KLEIN
L.D. LANDAU
(by E.M. Lifshitz)

World Scientific
Singapore • New Jersey • London • Hong Kong

Published by

World Scientific Publishing Co. Pte. Ltd.,
P O Box 128, Farrer Road, Singapore 9128
USA office: 687 Hartwell Street, Teaneck, NJ 07666
UK office: 73 Lynton Mead, Totteridge, London N20 8DH

FROM A LIFE OF PHYSICS

ISBN 9971-50-937-7

Printed in Singapore by JBW Printers & Binders Pte. Ltd.

FOREWORD: TWENTY-ONE YEARS AFTER

It is nearly 21 years since the organisation of the lectures on their lives of Physics which are reprinted in the accompanying volume. My first thought is sadness at the number of those who lectured then and who have departed from us (these include Professors Werner Heisenberg, Oscar Klein, Paul Adrian Maurice Dirac and Eugene M. Lifshitz).

Hans Bethe and Eugene Wigner are fortunately with us. May they live very long and active lives, to act as inspiration for us all, as they always have in the past.

These lectures were given within the context of a month-long Conference on Contemporary Physics which assembled to review the whole subject of Physics. It was natural to ask those who had created the subject to speak of their lives of Physics and to list the problems which they felt were still unsolved.

One of the problems which falls in this category is that of the interpretation of Quantum Mechanics − not only for individual particles but also for the universe as a whole. Some of the problems − on the other hand − particularly the problem of infinities which was Dirac's major worry − seem to be near to some sort of resolution by the heterotic superstring, at least in ten dimensions. (Fundamental

Particle Physics has moved in other directions which are concerned with the unification of fundamental forces, on the model of the electroweak unification; leading on to the Grand Unification Theories and culminating in Superstrings which promise to be Theories of Everything.) With the synthesis which phase transitions in the early universe provide us, the Standard Model of Particle Physics and Early Cosmology appear to have converged so that instead of two disciplines, we have just one scientific discipline. Naturally, other new problems have arisen, for example, High-Temperature Superconductivity, a subject ill-understood at the present moment. Large Scale Cosmology and Dark Matter present other unresolved issues.

But I wonder whether there is not a more significant change in the climate of Physics. We, at the Centre, are particularly sensitive to the recent emphasis on the role of Physics in development. In fact, we are now contemplating three new Centres to be set up on the lines of the International Centre for Theoretical Physics. These new Centres will cater particularly to the developing country needs and will consist of 1) an International Centre for High Technology and New Materials, 2) an International Centre for Earth Sciences and the Environment, and 3) an International Centre for Chemistry, Pure and Applied. Unlike the present Centre, these new Centres will undertake also experimental research and training. (The entire complex, consisting of the old and the new centres, will be called the International Centre for Science.)

I wonder whether a Conference organised by the new International Centre for Science will be slanted towards fundamental sciences only, as was the case in 1968 when there was the sense of gratitude and adulation which everyone felt towards the great men of Physics still amongst us.

It is in this context that the appropriateness of this volume being reprinted under the auspices of World Scientific Publishing in Singapore becomes apparent. Singapore was a developing country till recently but has, through its own efforts, based on a purposeful utilisation of modern science and technology, acquired a different status now. I wish to thank Professor K. K. Phua for insisting that this volume must be revised.

Abdus Salam
Trieste, 16 February 1989

CONTENTS

ENERGY ON EARTH AND IN THE STARS

Hans A. Bethe

Professor Salam opened the series of lectures by saying:

One of the purposes of the Symposium now in progress, as we conceived it, was to try to bridge among the many who are here the generation gap; to bring nearer to us the men who have created our subject and whom we have all admired from a distance. So a Life of Physics series was conceived to run coincidently with the Symposium. It provides the opportunity for some of our Grand Old Men to tell us the milieu of *physics* they have helped to create, illustrating it through their own work. We were unfortunate that Professor Weisskopf, who was to have been our first speaker, was prevented by illness from coming. So the starting of the series was delayed.

Tonight we have the privilege and honour of welcoming Professor Hans Bethe. Professor Robert Marshak has kindly agreed to take the chair. Professor Marshak, no stranger to Trieste, was one of the three, not old, but very wise men who selected Trieste in preference to other sites in 1963 for the location of the Centre. You will all agree what a wise choice his committee, which consisted of Professor Van Hove and Professor Tiomno besides Professor Marshak, made. Marshak is a member of the Scientific Council of the Centre.

R.E. Marshak:

Hans Bethe was born in 1906 in Strasbourg, Alsace-Lorraine. His father was a well-known physiologist at the University while his mother was a musician and writer of children's plays. Young Hans attended Goethe Gymnasium, a classical public school, and then left for the University of Frankfurt from which he was graduated in 1926. At the tender age of 22, Bethe gained his Ph.D. from the University of Munich under the famous theoretical physicist, Arnold Sommerfeld. Sommerfeld introduced Bethe to the excitement of modern physics and Enrico Fermi, with whom Bethe worked as a Rockefeller Foundation Fellow in Rome (1930–32), completed Bethe's early training. By 1935, Hans Bethe was settled at Cornell University — a refugee from Nazi Germany — and he has worked there ever since except for his years of war research and numerous visiting appointments.

I first came to Cornell in 1937 to do my graduate work under Professor Bethe. I got there rather accidentally by attending a conference on solid state physics which Bethe had organized. Bethe's versatility was already clear when I took up residence as a graduate student since he had just completed his monumental articles on nuclear physics. Within one year, Bethe's scientific activity had moved into the completely alien territory of astrophysics in the form of the epoch-making paper entitled "Energy Production in Stars". This paper was an outgrowth of a small theoretical conference at George Washington University (Washington, D.C.) organized by George Gamow and Edward Teller. Hans Bethe came back from this conference in the spring of 1938, greatly challenged by the problem of the origin of stellar energy. After several months, he had examined every conceivable nuclear reaction which might produce substantial amounts of energy under stellar conditions and had reached the conclusion that the carbon cycle and the proton-proton series of reactions were the two major sources of energy for the common main sequence stars. Bethe's profound analysis of the thermonuclear processes operating in the stars led to his winning the A. Cressy Morrison Astronomical Prize of the New York Academy of Sciences in 1938, the Draper Medal of the National Academy of Sciences in 1947, the Eddington Medal of the Royal Astronomical Society in 1963 and, finally, the Nobel Prize in 1967. Much more could be said about the importance of Hans Bethe's contribution to modern astrophysics but I merely point out that the tremendous effort underway at present to achieve self-sustaining fusion reactions in contained plasmas is a heroic attempt to duplicate on earth the thermonuclear processes so thoroughly analyzed by Bethe for the stars.

When World War II broke out, Bethe demonstrated his ability to apply his physical knowledge to problems of practical importance when the need arises. He first applied his knowledge of electromagnetic theory to radar problems at the M.I.T. Radiation Laboratory and concluded his war work as the Head of the Theoretical Division of the Los Alamos Laboratory. I served under Bethe at both laboratories and can testify to the tremendous energy and understanding which he brought to bear on the most diverse applied problems.

After the end of the war, in the summer of 1946, Bethe and I were both consultants at the General Electric Research Laboratories in Schenectady, trying to communicate to this active laboratory the new wonders of "atomic energy". Some of the "youngsters" we educated at that time in nuclear reactor theory were Harvey Brooks and Henry Hurwitz, destined to become scientific leaders in their own right. And a year later our paths crossed again at the famous Shelter Island Conference where Bethe was inspired to work out the non-relativistic theory of the Lamb shift.

A couple of years ago I decided to arrange for a little volume in honor of Bethe's 60th birthday which would try to reflect and recapture the broad and versatile contributions which he has made to almost every branch of physics. The response was so overwhelming and the coverage of physics and astronomy so complete that I was compelled to use Sam Goudsmit's classification for the *Physical Review* to order properly the various articles in the book!

PLEASURE FROM PHYSICS

Hans A. Bethe:

When Salam asked me to give this talk he wrote that I should talk about the things in physics which I particularly enjoyed. That is exactly what I am going to do.

The first work which I enjoyed very much was the paper about the stopping power of matter. The Born theory of atomic collisions had just appeared and Elsasser had applied it to the scattering of electrons from hydrogen atoms, both elastic and inelastic. Out of that calculation there came rather long and unwieldy formulae, getting worse the higher the quantum number of the excited state, and there was no way to foresee how terrible these formulae would be when you came to the excitation of states in the continuous spectrum. So I thought this was not a way to make a living and one should be able to do it more simply. Essentially I did two things in this paper, one of which was to discover Poisson's equation. The Born approximation tells you that the scattering amplitude is given by

$$\int V(\mathbf{r}) \, e^{i\mathbf{q}\cdot\mathbf{r}} \, d^3r.$$

The potential $V(\mathbf{r})$, in the case of elastic scattering, is of course just that made by the charge distribution of the atom, and $\mathbf{q}$ is the change of momentum. As said, I discovered Laplace's equation, by saying that the potential after all is connected with the charge density and thereby the Born approximation result can be transformed into

$$q^{-2} \int \rho(\mathbf{r}) \, e^{i\mathbf{q}\cdot\mathbf{r}} \, d^3r.$$

This, of course, gives a very much simpler relation to the actual properties of the atom, since the density is directly given by the wave function.

The second trick I used in the paper was a sum rule. Of course this was not the first time that sum rules were used. The Kuhn-Reiche-Thomas sum rule had in fact been the basis of quantum mechanics, a couple of years earlier, but sum rules had not yet been thoroughly exploited for the simplification of results in quantum mechanics. Now, it is simple to get a sum rule for the total cross section. In the case of inelastic scattering, of course, $\rho(r)$ is replaced by

$$\Psi_n^*(\mathbf{r}) \, \Psi_0(\mathbf{r}),$$

the final state times the initial state wave function. If you sum the cross section over all the excited states you essentially get the absolute square of the operator

$$\sum_i e^{i\mathbf{q}\cdot\mathbf{r}_i},$$

which is unity in the case of hydrogen; so you can tell the total cross section for excitation of all states.

EXPRESSING ENERGY LOSS

The sum rule which was not quite so trivial was that which corresponds to the energy loss of a particle. The quantity I have discussed is the scattered amplitude $F_n(g)$, its absolute square is the cross section. If you now multiply this square by the energy loss of the incident particle and sum over all excited states, this gives you the probability of scattering, multiplied by the average energy loss; it essentially gives you the energy loss per atom traversed. Fortunately, I found that there is a very simple sum rule for this: in fact it is even simpler than the one giving the total cross section. If q is very small, the matrix element reduces to q times the dipole moment, and hence the energy loss sum becomes just the Thomas-Reiche-Kuhn rule, which gives the number of electrons. But the interesting fact was that also in the case where q is not small, this energy loss sum also gives just the number of electrons. This, I still think, is very remarkable and it certainly was very pleasant to me at the time. So I was able, starting from the theory of Born, to arrive at a closed expression for the energy loss, per unit length of path, for the charged particle traversing matter.

I should mention that I used this paper to become a privatdozent in Munich. As you know, this is a special ritual in Germany. Having a doctor's degree does not entitle you to be a teacher at a German University, you have to pass a second examination which consists of writing an acceptable paper and presenting a certain number of "thesis", i.e., one has to make a number of claims and the whole faculty of science can come and attack the candidate and prove that his claims are wrong. Of course, this is only a formality and the faculty is very gentle!

After a year or two, I met Professor Blackett at Cambridge and he told me: "Now look here, you have made a theory of the energy loss of charged particles, but your qualitative results are no good to me, I really want to know this energy loss quantitatively, and I want to know it so accurately that I can measure the range of a particle and from that range deduce the energy of the particle." At that time, it was quite difficult to have a good enough electrical apparatus to measure the energy of a particle by electric and magnetic deflection. The range was then the most accessible measurement of the energy of a particle. So Blackett said, "Well, there is a paper by Mr. Duncanson who has calculated the range on the basis of the old Bohr theory which we know is not very good, why don't you do it again on the basis of quantum theory?" So I was led into generalizing my theory to the case of complex atoms. This introduced the average excitation potential of the atom which was still an empirical constant which had to be determined from the measurement of one range for a certain energy; from that, the range-energy relation for all energies could be deduced. This scheme worked surprisingly well and kept me busy for many years refining it, that is, my students and I put in corrections for the fact that some of the electrons have strong binding energy.

Then, a few years later, Moller's theory of relativistic collisions came along. One of the things I did was to apply this to the stopping power problem, and I found that it worked out pretty much in the same way as in the non-relativistic case. I found out just a few days ago that almost at the same time, Oppenheimer did the same calculation. Well this much for the stopping power; it was a most satisfactory field because you could come from first principles to something which could really be compared with experiment, and could actually be helpful to the experimentalist in his interpretation.

SOLVED IN THE SUBWAY

The most satisfactory period of my life was in the 1930's, the development of nuclear physics. This started in Manchester when I lived and worked together with Peierls and we were both very interested in the deuteron. This was being investigated experimentally at the time by Chadwick and Goldhaber. We considered especially the relation of the binding energy of the deuteron to the scattering of neutrons by protons. We found at that time (this was mostly Peierls' suggestion) a very close relation between the scattering cross section and the deuteron binding energy, essentially the scattering cross section was

$$\sigma = \frac{4\pi\hbar^2}{M} \cdot \frac{1}{E + \epsilon},$$

i.e., a constant, divided by the energy of the system of the neutron and the proton in the centre-of-mass system, plus the binding energy. Well our theory was very nice, but it did not agree with experiment! The better the experiments the worse the discrepancy. Finally, the solution to this was told to me in 1935 in a subway train in New York by Eugene Wigner. I don't know, I must have been able to hear much better than I do now, which I know is true, and probably Wigner spoke louder than he does now! At any rate, I was able to hear him in the subway, and he said, "Now look here, all that the deuteron tells you is the interaction of neutrons and protons in the triplet state. How do you know how they interact in the singlet state? Probably they interact quite differently." So he solved this problem between Columbia University and Pennsylvania station, but he never published it. I published it, giving him credit for it, in one of the three *Reviews of Modern Physics* articles, but I am very sorry that he never wrote it up.

ATOMIC MASSES

In addition to the binding energy and the scattering, we also did the photoelectric disintegration of the deuteron which just at that time had been observed by Chadwick and Goldhaber. It was Goldhaber who led me to the second subject in nuclear physics, by pointing out that the atomic masses were in a terrible mess. This was especially true of ^{9}Be. According to the published masses, ^{9}Be should really not exist, because it had a mass bigger than a neutron plus two alpha particles! So the next thing I did was to look at the atomic masses and try to use as much as possible, energies which came from nuclear disintegrations, which by that time had been observed in large numbers, and to take only those mass spectrograph measurements which seemed very reliable. Essentially only the measurements made at Cambridge and at Harvard fell into this category. Combining these with disintegration data I was able to construct a table of atomic masses. This was about my first work at Cornell in 1935, and this table no longer gave to ^{9}Be too large a mass, but made it perfectly stable. The deuteron also was in this category. The mass spectrograph data had indicated that the deuteron also ought to disintegrate spontaneously into a neutron and a proton. In the course of time, the mass spectrograph people changed their data and confirmed that really the nuclear disintegration values were the correct ones.

At Cornell I was in contact with experimentalists. We had a cyclotron, the second one ever built, which had been built by Livingston who had been a very close collaborator of Ernest Lawrence. Our cyclotron was very small because the department could only afford, I think, two or three thousand dollars to build it. For many years we were very proud that we had the smallest cyclotron in operation. We claimed that per dollar and per kilowatt, we have produced more research than any other cyclotron.

Well I found that the experimentalists didn't know much about nuclear physics and that it was very laborious to explain the same things to one experimenter after another. So I decided it would be much easier to write it down, and so I wrote the three articles in the *Reviews of Modern Physics* which Segrè called the "Mattoncino", (a small rock I think). In these I put down most of the things that were then known about nuclear physics. This was lots of fun; I had two collaborators, Konopinski and M.E. Rose who sat in an office together and did all the calculations that I needed done. We would only interrupt this work when Konopinski came to tell me it was time to eat; so we would go to eat, as Marshak described it. We were able to fill in a great number of gaps in the then existing knowledge, like excitation functions, something about the diffusion of neutrons, and quite a bit about binding energies of light nuclei. Also at that time people began to play with the shell model. The shell model worked fine up to 40 Ca but not beyond.

It was very satisfactory to describe all this in the *Reviews* article, and it was even more satisfactory that many experimentalists were interested in confirming some of the theories that we had put down. There was an especially close collaboration with the University of Rochester which had a bigger cyclotron, and therefore could test quite a number of the theories of excitation functions that we had developed. At about that time, Bohr invented the compound nucleus, Breit and Wigner found the dispersion formula, and in connection with the resonances predicted by these theories I did the only experiment which I ever had done personally. There was an experimental graduate student at Cornell who measured the radioactivity produced in silver by slow neutrons and our problem was to prove that the energy of these neutrons was different from the thermal energy. We did this with the help of the boron absorption which was simultaneously used for the same purpose in England by Moon and others. In this case, I really sat at night in one of the laboratories and counted the number of counts made by these neutrons.

SOLID STATE

Marshak has mentioned solid state as the reason why I was not hired by Pauli. I worked on solid state but I must say at the time it was a far less satisfactory pursuit than nuclear physics. It was really much too early to do solid state seriously. My ambition at the time was to calculate such things as the shape of the Fermi surface for the electrons in silver or at least in sodium, and then to have some experimental confirmation for this. By now we know the shapes of the Fermi surfaces of these substances, but at that time there was absolutely no way of doing it, all you could measure was conductivity and a few thermoelectric and magnetic effects, gross numbers which certainly would never give you any information of the kind I wanted. Wigner and Seitz, at the end of the period I worked on solid state physics, invented their very powerful method to determine theoretically the allowed bands of electrons in metals. This was obviously the way to do it, but at least I lacked the mathematical power to put this method really to work and to get out of it the information which you need to get, let us say, the shape of the Fermi surface.

There were two other things which I did on solid state theory. One was the splitting of atomic energy levels when the atom is inserted into a crystal, into a site of given symmetry. I did that essentially only because I had studied a book on group theory, and you can't really understand something unless you apply it and work with it yourself.

So since Wigner had done all the really important things with group theory, I thought the only thing that remains to be done is to take an atom in a crystal of various symmetries and see how the energy levels will look there. I am told that people have used that paper, but I have never seen what has come out of it.

Another thing in solid state theory was the order and disorder in alloys. This problem was suggested to me by Sir Lawrence Bragg, while I was in Manchester at the same period when I also did the deuteron with Peierls. What I did was to put just a very slight amount of respectability into the theory of Bragg and Williams on the way order and disorder come about.

NUCLEAR REACTIONS IN THE STARS

Marshak has mentioned one of the side-lines which grew out of nuclear physics and out of the theory of nuclear reactions, namely the energy production in stars. This was one of the most satisfactory things to work on. As Marshak mentioned, there was a little Conference in Washington, organized by the Department of Terrestrial Magnetism which is not a Government department but a part of the Carnegie Foundation. At this Conference the astrophysicists told some of us physicists what stars are about, how they are made, what distribution of density and pressure they have and so on, and then they ended up with the question where does the energy come from? Everybody of course agreed that the energy must come from nuclear reactions but what nuclear reactions? They were searching at the time for too much, viz. they were trying at the time to solve simultaneously the problem of the buildup of elements and the problem of production of energy in the stars. It was just the removal of this coupling that made it possible to solve the problem.

I found the carbon cycle in a very systematic way. Weizsacker had first suggested that perhaps the fundamental reaction in stars is the simplest one that you can think of, namely,

$$H + H = D + \epsilon^+ + \nu.$$

This, of course, is a reaction which is exceedingly improbable because it involves beta decay, but in stars we have almost unlimited time plus very high densities and high enough temperature to overcome the potential barrier. And so it turned out that indeed this reaction gives just about the right rate of energy production in the sun, and today it is believed to be the main reaction in the sun.

However, we were told by our astrophysicist friends that there are stars which are much more brilliant than the sun and which yet have an internal temperature only slightly higher than that of the sun. And this was the puzzle, how could these stars give such enormous rates of energy production? Now the proton-proton reaction has a very weak energy dependence because at the temperature which exists at the centre of the sun, the potential barrier is quite easily penetrated. The reaction rate behaves approximately as the fourth power of the temperature and that is not nearly enough to explain the brilliance of, let us say, Sirius A and more brilliant stars. So I had to look for a reaction which involved atoms with higher potential barriers. So I went systematically through the periodic table but everything gave nonsense because whatever atom I used, lithium, beryllium, etc., would be destroyed in the reaction, and there was very little of these substances anyway, as we know from abundances both on earth and in the stars. So these elements could not possibly give the energy production for the length of time that the universe has been functioning. Finally I got to carbon, and as you all know, in the case of carbon the reaction works out beautifully. One goes through six reactions, and at the end one comes back to carbon. In the process, one has made four hydrogen atoms into one helium. This theory of course, was not made on the railway train from Washington to Ithaca (as was claimed in Gamow's book). It didn't take very long, it took about six weeks, but not even the Trans-Siberian railroad took that long for its journey.

THE LAMB SHIFT

There was however, one paper which I did do on a train, and that was the Lamb shift. In 1947, Lamb and Rutherford had discovered the shift of the $2s$-state of hydrogen upward in energy. We had a very beautiful conference on Shelter Island in which these and other experimental results were discussed, together with the state of theoretical physics which was plagued by the infinities of self-energy. Of course, people had struggled with self-energy infinities for a long time, in particular Oppenheimer and Kramers. Kramers suggested that what one really ought to do was to renormalize the mass of the electron, taking into account its interaction with its own electromagnetic field. Then only those parts of the self-energy which are not contained in the mass of the particle would be observable and amenable to experimentation.

I found this suggestion very interesting and I thought that it ought to be possible to get Lamb's result by applying the idea of Kramers. So on the train from Shelter Island to the General Electric Company in Schenectady which was also mentioned by Marshak, I wrote down some elementary equations of radiation interaction and found out that the effect on the $2s$-state or any state of hydrogen would involve the logarithm of the energy. Inside the logarithm, the numerator was some energy which I did not know, while in the denominator, there was something like the binding energy of the electron in hydrogen. So without doing anything about it, this expression would depend only logarithmically on the upper limit of the energy of the interacting quantum. This sounded very hopeful: I had used entirely nonrelativistic theory, and we know that, for instance, the electromagnetic mass diverges linearly if one takes the non-relativistic theory, but diverges only logarithmically, if one takes relativistic theory.

It was reasonable to conclude that relativistic theory would gain us one power which meant that the logarithmic divergence would be replaced by convergence. Stupidly or boldly, I just assumed that the higher energy was mc^2, and with this assumption, I got about the right answer. Of course, I was afraid that I might have made a mistake by a factor of 2 in writing down the interaction between the electron and the electromagnetic field — after all one cannot remember factors of two on a train. So the next morning, as early as I could, I looked for Heitler's book in the General Electric library, and found that I had not made a mistake. Indeed I got a result of about a thousand megacycles which was about the right answer.

However, I was not sufficiently powerful to conclude this work and to get the relativistic theory; this was done by people who knew a great deal more, Schwinger, Feynman, Tomonaga and a few others. They brought this problem to a conclusion and showed that quantum electrodynamics can be renormalized in a consistent way. As you know, this gives finite answers for every order of perturbation theory, and in particular, about the right answer for the Lamb shift. I am told the answer is still not quite right, it is still wrong by two-tenths of a megacycle and nobody knows what that discrepancy is due to.

WARTIME WORK

I have mentioned so far only pure physics work. Just for balance, I should mention that I have had a great deal of satisfaction from applied work. This started with the work on radar at the Radiation Laboratory at M.I.T., which Marshak has mentioned. It continued with Los Alamos, where we had the rather gruesome task of designing the atomic bomb. But this problem really had a great scientific challenge; perhaps the greatest challenge was the need for combining knowledge in very different fields. We had to know nuclear physics, of course. A lot of predictions of cross sections of various reactions at various energies required a knowledge of nuclear physics. Weisskopf was the great master in this and was known as the Los Alamos oracle. Then we had to know neutron diffusion theory on which Wigner had done the most fundamental work.

Then there was hydrodynamics. It was decided at a very early time at Los Alamos that the best way to bring a critical mass of material together was to have a spherical shell of uranium, to surround this with another spherical shell of explosive, and then to detonate that explosive in such a way as to set up an ingoing detonation wave. This would simultaneously hit all points of the spherical shell of uranium and implode it. This implosion, we thought, would not only assemble the spherical shell into one compact sphere and thereby get us beyond the critical mass very rapidly, but in addition would compress the metal in the sphere well beyond the normal density of metal.

To estimate this effect we had to guess, first of all, an equation of state under conditions that nobody on earth had ever seen, namely a few million atmospheres pressure which can be produced in a converging shock wave. Next, we had to calculate how the uranium would move under the influence of the high explosive impulse taking into account its equation of state. We used, I think for the first time to that extent, computers to do these complicated calculations and we found an answer. It was also challenging to check and to supplement the answer by analytical work and this we also did. This applied work, combining so many disciplines was one of the really interesting pieces of work in physics which I have done.

MATERIALS FOR SPACE VEHICLES

Since then, I have constantly been doing some applied work. Let me mention just one, namely the re-entry of space vehicles into the atmosphere. Because of the tremendous speeds there is a tremendous heating, and one has to know what material to choose for the heat shield. It was suggested to use some glassy material, quartz for instance. Now the trouble with glass is that at a certain temperature it begins to flow. So the problem was: would this glassy material flow away and only absorb the heat necessary to raise its temperature to melting? In this case, the whole idea was no good because the glass would absorb very little energy, something like a thousand calories a gram. To absorb more energy one needs evaporation of the material. So the question was, would the material flow away before it evaporated? We knew the shear forces (they can be calculated from the aerodynamics), but the problem of the glassy layer with a high but still finite viscosity, under influence of the shear forces and high, variable temperature was too much for orthodox aerodynamics. My contribution was to reverse the usual assumption which aerodynamicists like, namely that viscosity is a constant. In this case, one had to assume that viscosity was the most rapidly variable function of position. This was very much against the aerodynamicists' instinct, but we solve the problem and it worked, it does evaporate.

NEW WORK ON NUCLEAR MATTER

Let me return to pure physics and to the problem I have worked on over the last thirteen years, namely nuclear matter. I came to nuclear matter mainly because the pace of high energy physics is too fast for me. By the time I have learnt the newest method of high energy physics, it is discarded, so this is no field for an old man, particularly for an old man who has also some other things to do in addition to research. So I found myself nuclear matter, and started out by trying to understand the theories of Brueckner. They were very obscure. I spent a sabbatical leave trying to understand what he meant; then wrote down what I thought he meant and he agreed that this was indeed what he meant. I also found a very able graduate student in Cambridge, Goldstone, who put the Brueckner theory on a sound basis. Goldstone went back to Feynman diagrams and proved Brueckner's theory step by step, showing that you can get a consistent set of approximations.

After this, the problem was to exploit the theory. First we wished to calculate results which we know, such as the binding energy of nuclear matter which is 16 millions volts per particle, and the density of nuclear matter. We still haven't finished that job, but we have come fairly close to the right result. Using a good interaction between two nucleons, we now find a binding energy of about 13.5 MeV. What the remaining 2.5 is due to, I don't know, it may well be due to a velocity dependence of nuclear forces, because we always assume that nuclear forces are static, simply because this is the simplest concrete assumption which you can make, not because I necessarily believe that it is so.

This calculation of the binding energy is merely to establish the method, to find that one can calculate some known quantity and get the right answer, but what one really wants to do, of course, is to calculate unknown quantities which cannot so easily be obtained experimentally. The problem which I have been busy with for the last three of four years, has been to take a first step toward finite nuclei, namely to consider a semi-infinite nucleus, that is one which has a surface. In many different ways, we have calculated the shape and the thickness of the nuclear surface, and I am happy to see that as of last Monday, my student who is doing this got a good answer. His answer agrees with the thickness of the surface as observed in electron scattering by the Stanford Group, about 2.4 fermi.

Earlier, we were able to get the correct surface energy, i.e. the constant in the Weizsacker formula which goes with the surface energy. We got for this an embarrassingly good result which agrees with experiment to the accuracy of experiment and of the theory. We are now trying to do this also for actual nuclei like ^{208}Pb which has different numbers of neutrons and protons, so the Coulomb and symmetry energy have to be put in. We have some indication that we get things like the wine-bottle shape of the charge distribution which seems to be well established experimentally by a combination of electron scattering and of the X-ray energy levels of muonic atoms.

Then of course we are busy with actual finite nuclei, trying to develop methods to calculate finite nuclei. We start either from the fundamental nuclear force, or maybe from an intermediate step (which is what I favour), namely from the results which we have already obtained for infinite nuclear matter of various density. From this we try to obtain a simple and reliable theory of finite nuclei to answer such questions as: such and such a nucleus should be more stable elongated, while another nucleus should be more stable in spherical shape, telling also the energy difference between different shapes and the energy levels, as much as possible from first principles. Much of this has already been done by Gerry Brown at Princeton who has done a lot about the interaction of nucleons in incomplete shells, starting from a realistic interaction between the nucleons.

These have been some of the things I have been interested in, some of the things which have given me pleasure, and I wish that all of you will have as much pleasure in doing physics as I have had.

METHODS IN THEORETICAL PHYSICS

Paul Adrian Maurice Dirac

I shall attempt to give you some idea of how a theoretical physicist works — how he sets about trying to get a better understanding of the laws of nature.

One can look back over the work that has been done in the past. In doing so one has the underlying hope at the back of one's mind that one may get some hints or learn some lessons that will be of value in dealing with present-day problems. The problems that we had to deal with in the past had fundamentally much in common with the present-day ones, and reviewing the successful methods of the past may give us some help for the present.

One can distinguish between two main procedures for a theoretical physicist. One of them is to work from the experimental basis. For this, one must keep in close touch with the experimental physicists. One reads about all the results they obtain and tries to fit them into a comprehensive and satisfying scheme.

The other procedure is to work from the mathematical basis. One examines and criticizes the existing theory. One tries to pin-point the faults in it and then tries to remove them. The difficulty here is to remove the faults without destroying the very great successes of the existing theory.

There are these two general procedures, but of course the distinction between them is not hard-and-fast. There are all grades of procedure between the extremes.

Which procedure one follows depends largely on the subject of study. For a subject about which very little is known, where one is breaking quite new ground, one is pretty well forced to follow the procedure based on experiment. In the beginning, for a new subject, one merely collects experimental evidence and classifies it.

For example, let us recall how our knowledge of the periodic system for the atoms was built up in the last century. To begin with, one simply collected the experimental facts and arranged them. As the system was built up one gradually acquired confidence in it, until eventually, when the system was nearly complete, one has sufficient confidence to be able to predict that, where there was a gap, a new atom would subsequently be discovered to fill the gap. These predictions all came true.

In recent times there has been a very similar situation for the new particles of high energy physics. They have been fitted into a system in which one has so much confidence that, where one finds a gap, one can predict that a particle will be discovered to fill it.

In any region of physics where very little is known, one must keep to the experimental basis if one is not to indulge in wild speculation that is almost certain to be wrong. I do not wish to condemn speculation altogether. It can be entertaining and may be indirectly useful even if it does turn out to be wrong. One should always keep an open mind receptive to new ideas, so one should not completely oppose speculation, but one must take care not to get too involved in it.

COSMOLOGICAL SPECULATION

One field of work in which there has been too much speculation is cosmology. There are very few hard facts to go on, but theoretical workers have been busy constructing various models for the universe, based on any assumptions that they fancy. These models are probably all wrong. It is usually assumed that the laws of nature have always been the same as they are now. There is no justification for this. The laws may be changing, and in particular quantities which are considered to be constants of nature may be varying with cosmological time. Such variations would completely upset the model makers.

With increasing knowledge of a subject, when one has a great deal of support to work from, one can go over more and more towards the mathematical procedure. One then has as one's underlying motivation the striving for mathematical beauty. Theoretical physicists accept the need for mathematical beauty as an act of faith. There is no compelling reason for it, but it has proved a very profitable objective in the past. For example, the main reason why the theory of relativity is so universally accepted is its mathematical beauty.

With the mathematical procedure there are two main methods that one may follow, (i) to remove inconsistencies and (ii) to unite theories that were previously disjoint.

SUCCESS THROUGH METHOD

There are many examples where the following of method (i) has led to brilliant success. Maxwell's investigation of an inconsistency in the electromagnetic equations of his time led to his introducing the displacement current, which led to the theory of electromagnetic waves. Planck's study of difficulties in the theory of black-body radiation led to his introduction of the quantum. Einstein noticed a difficulty in the theory of an atom in equilibrium in black-body radiation and was led to introduce stimulated emission, which has led to the modern lasers. But the supreme example is Einstein's discovery of his law of gravitation, which came from the need to reconcile Newtonian gravitation with special relativity.

In practice, method (ii) has not proved very fruitful. One would think that the gravitational and electromagnetic fields, the two long-range fields known in physics, should be closely connected, but Einstein spent many years trying to unify them, without success. It seems that a direct attempt to unify disjoint theories, where there is no definite inconsistency to work from, is usually too difficult, and if success does ultimately come, it will come in an indirect way.

Whether one follows the experimental or the mathematical procedure depends largely on the subject of study, but not entirely so. It also depends on the man. This is illustrated by the discovery of quantum mechanics.

Two men are involved, Heisenberg and Schrödinger. Heisenberg was working from the experimental basis, using the results of spectroscopy, which by 1925 had accumulated an enormous amount of data. Much of this was not useful, but some was, for example the relative intensities of the lines of a multiplet. It was Heisenberg's genius that he was able to pick out the important things from the great wealth of information and arrange them in a natural scheme. He was thus led to matrices.

Schrödinger's approach was quite different. He worked from the mathematical basis. He was not well informed about the latest spectroscopic results, like Heisenberg was, but had the idea at the back of his mind that spectral frequencies should be fixed by eigenvalue equations, something like those that fix the frequencies of systems of vibrating springs. He had this idea for a long time, and was eventually able to find the right equation, in an indirect way.

IMPACT OF RELATIVITY

In order to understand the atmosphere in which theoretical physicists were then working, one must appreciate the enormous influence of relativity. Relativity had burst into the world of scientific thought with a tremendous impact, at the end of a long and difficult war. Everyone wanted to get away from the strain of war and eagerly seized on the new mode of thought and new philosophy. The excitement was quite unprecedented in the history of science.

Against this background of excitement, physicists were trying to understand the mystery of the stability of atoms. Schrödinger, like everyone else, was caught up by the new ideas, and so he tried to set up a quantum mechanics within the framework of relativity. Everything had to be expressed in terms of vectors and tensors in space-time. This was unfortunate, as the time was not ripe for a relativistic quantum mechanics, and Schrödinger's discovery was delayed in consequence.

Schrödinger was working from a beautiful idea of de Broglie connecting waves and particles in a relativistic way. De Broglie's idea applied only to free particles, and Schrödinger tried to generalize it to an electron bound in an atom. Eventually he succeeded, keeping within the relativistic framework. But when he applied his theory to the hydrogen atom, he found it did not agree with experiment. The discrepancy was due to his not having taken the spin of the electron into account. It was not then known. Schrödinger subsequently noticed that his theory was correct in non-relativistic approximation, and he had to reconcile himself to publishing this degraded version of his work, which he did after some months' delay.

The moral of this story is that one should not try to accomplish too much in one stage. One should separate the difficulties in physics one from another as far as possible, and then dispose of them one by one.

Heisenberg and Schrödinger gave us two forms of quantum mechanics, which were soon found to be equivalent. They provided two pictures, with a certain mathematical transformation connecting them.

I joined in the early work on quantum mechanics, following the procedure based on mathematics, with a very abstract point of view. I took the non-commutative algebra which was suggested by Heisenberg's matrices as the main feature for a new dynamics, and examined how classical dynamics could be adapted to fit in with it. Other people were working on the subject from various points of view, and we all obtained equivalent results, at about the same time.

FRUITFUL RELAXATION

I would like to mention that I found the best ideas usually came, not when one was actively striving for them, but when one was in a more relaxed state. Professor Bloch has told us how he got ideas on railway trains and often worked them out before the end of the journey. It was not like that with me. I used to take long solitary walks on Sundays, during which I tended to review the current situation in a leisurely way. Such occasions often proved fruitful, even though, (or perhaps because,) the primary purpose of the walk was relaxation and not research.

It was on one of these occasions that the possibility occurred to me of a connection between commutators and Poisson brackets, I did not then know very well what a Poisson bracket was, so was very uncertain of the connection. On getting home I found I did not have any book explaining Poisson brackets, so I had to wait impatiently for the libraries to open the following morning before I could verify the idea.

With the development of quantum mechanics one had a new situation in theoretical physics. The basic equations, Heisenberg's equations of motion, the commutation relations and Schrödinger's wave equation were discovered without their physical interpretation being known. With non-commutation of the dynamical variables, the direct interpretation that one was used to in classical mechanics was not possible, and it became a problem to find the precise meaning and mode of application of the new equations.

This problem was not solved by a direct attack. People first studied examples, such as the non-relativistic hydrogen atom and

Compton scattering, and found special methods that worked for these examples. One gradually generalized, and after a few years the complete understanding of the theory was evolved as we know it today, with Heisenberg's principle of uncertainty and the general statistical interpretation of the wave function.

The early rapid progress of quantum mechanics was made in a non-relativistic setting, but of course people were not happy with this situation. A relativistic theory for a single electron was set up, namely Schrödinger's original equation, which was rediscovered by Klein and Gordon and is known by their name, but its interpretation was not consistent with the general statistical interpretation of quantum mechanics.

FROM TENSORS TO SPINORS

As relativity was then understood, all relativistic theories had to be expressible in tensor form. On this basis one could not do better than the Klein-Gordon theory. Most physicists were content with the Klein-Gordon theory as the best possible relativistic quantum theory for an electron, but I was always dissatisfied with the discrepancy between it and general principles, and continually worried over it till I found the solution.

Tensors are inadequate and one has to get away from them, introducing two-valued quantities, now called spinors. Those people who were too familiar with tensors were not fitted to get away from them and think up something more general, and I was able to do so only because I was more attached to the general principles of quantum mechanics than to tensors. Eddington was very surprised when he saw the possibility of departing from tensors. One should aways guard against getting too attached to one particular line of thought.

The introduction of spinors provided a relativistic theory in agreement with the general principles of quantum mechanics, and also accounted for the spin of the electron, although this was not the original intention of the work. But then a new problem appeared, that of negative energies. The theory gives symmetry between positive and negative energies, while only positive energies occur in nature.

As frequently happens with the mathematical procedure in research, the solving of one difficulty leads to another. You may think that no real progress is then made, but this is not so, because the second difficulty is more remote than the first. It may be that the second difficulty was really there all the time, and was only brought into prominence by the removal of the first.

This was the case with the negative energy difficulty. All relativistic theories give symmetry between positive and negative energies, but previously this difficulty had been overshadowed by more crude imperfections in the theory.

The difficulty is removed by the assumption that in the vacuum all the negative energy states are filled. One is then led to a theory of positrons together with electrons. Our knowledge is thereby advanced one stage, but again a new difficulty appears, this time connected with the interaction between an electron and the electromagnetic field.

When one writes down the equations that one believes should describe this interaction accurately and tries to solve them, one gets divergent integrals for quantities that ought to be finite. Again this difficulty was really present all the time, lying dormant in the theory, and only now becoming the dominant one.

ON THE WRONG TRACK?

If one deals classically with point electrons interacting with the electromagnetic field, one finds difficulties connected with the singularities in the field. People have been aware of these difficulties from the time of Lorentz, who first worked out the equations of motion for an electron. In the early days of the quantum mechanics of Heisenberg and Schrödinger, people thought these difficulties would be swept away by the new mechanics. It now become clear that these hopes would not be fulfilled. The difficulties reappear in the divergencies of quantum electrodynamics, the quantum theory of the interaction of electrons and the electromagnetic field. They are modified somewhat by the infinities associated with the sea of negative-energy electrons, but they stand out as the dominant problem.

The difficulty of the divergencies proved to be a very bad one. No progress was made for twenty years. Then a development came, initiated by Lamb's discovery and explanation of the Lamb shift, which fundamentally changed the character of theoretical physics. It involved setting up rules for discarding the infinities, rules which are precise, so as to leave well-defined residues that can be compared with experiment. But still one is using working rules and not regular mathematics.

Most theoretical physicists nowadays appear to be satisfied with this situation, but I am not. I believe that theoretical physics has gone on the wrong track with such developments and one should not be complacent about it. There is some similarity between this situation and the one in 1927, when most physicists were satisfied with the Klein-Gordon equation and did not let themselves be bothered by the negative probabilities that it entailed.

We must realize that there is something radically wrong when we have to discard infinities from our equations, and we must hang on to the basic ideas of logic at all costs. Worrying over this point may lead to an important advance. Quantum electrodynamics is the domain of physics that we know most about, and presumably it will have to be put in order before we can hope to make any fundamental progress with the other field theories, although these will continue to develop on the experimental basis.

Let us see what can be done with putting the present quantum electrodynamics on a logical footing. We must keep to the standard practice of neglecting only quantities which one can believe to be small, even though the grounds for this belief may be rather shaky.

In order to handle infinities, we must refer to a process of cut-off. We must do this in mathematics whenever we have a series or an integral which is not absolutely convergent. When we have introduced a cut-off, we may proceed to make it more and more remote and go to a limit, which then depends on the method of cut-off. Alternatively, we may keep the cut-off finite. In the latter case, we must find quantities that are insensitive to the cut-off.

The divergencies in quantum electrodynamics come from the high-energy terms in the energy of interaction between the particles and the field. The cut-off thus involves introducing an energy, g say, beyond which the interaction energy terms are omitted. It is found that we cannot make g tend to infinity without destroying the possibility of solving the equations logically. We have to keep a finite cut-off.

The relativistic invariance of the theory is then destroyed. This is a pity, but it is a lesser evil than a departure from logic would be. It results in a theory which cannot be valid for high-energy processes, processes involving energies comparable with g, but we may still hope that it will be a good approximation for low-energy processes.

On physical grounds we should expect to have to take g to be of the order of a few hundred MeV, as this is the region where quantum electrodynamics ceases to be a self-contained subject and the other particles of physics begin to play a role. This value of g is satisfactory for the theory.

Working with a finite cut-off, we have to search for quantities which are not sensitive to the precise mode and value of the cut-off. We then find that the Schrödinger picture is not a suitable one. Solutions of the Schrödinger equation, even the one describing the vacuum state, are very sensitive to the cut-off. But there are some calculations that one can carry out in the Heisenberg picture that lead to results insensitive to the cut-off.

One can deduce in this way the Lamb shift and the anomalous magnetic moment of the electron. The results are the same as those obtained some twenty years ago by the method of working rules with discard of infinities. But now the result can be obtained by a logical process, following standard mathematics in which only small quantities are neglected.

As we cannot now use the Schrödinger picture, we cannot use the regular physical interpretation of quantum mechanics involving the square of the modulus of the wave function. We have to feel our way towards a new physical interpretation which can be used with the Heisenberg picture. The situation for quantum electrodynamics is rather like that for elementary quantum mechanics in the early days when we had the equations of motions but no general physical interpretation.

A feature of the calculations leading to the Lamb shift and anomalous magnetic moment should be noted. One finds that the parameters m and e denoting the mass and charge of the electron in the starting equations are not the same as the observed values for these quantities. If we keep the symbols m and e to denote the observed values, we have to replace the m and e in the starting equations by $m + \delta m$ and $e + \delta e$, where δm and δe are small corrections which can be calculated. This procedure is known as renormalization.

DIFFICULTY IN QUANTUM ELECTRODYNAMICS

Such a change in the starting equations is permitted. We can take any starting equations we like, and then develop the theory by making deductions from them. You might think that the work of the theoretical physicist is easy if he can make any starting assumptions he likes, but the difficulty arises because he needs the same starting assumptions for all the applications of the theory. This very strongly restricts his freedom. Renormalization is permitted because it is a simple change which can be applied universally whenever one has charged particles interacting with the electromagnetic field.

There is a serious difficulty still remaining in quantum electrodynamics, connected with the self-energy of the photon. It will have to be dealt with by some further change in the starting equations, of a more complicated kind than renormalization.

The ultimate goal is to obtain suitable starting equations from which the whole of atomic physics can be deduced. We are still far from it. One way of proceeding towards it is first to perfect the theory of low-energy physics, which is quantum electrodynamics, and then try to extend it to higher and higher energies. However, the present quantum electrodynamics does not conform to the high standard of mathematical beauty that one would expect for a fundamental physical theory, and leads one to suspect that a drastic alteration of basic ideas is still needed.

THEORY, CRITICISM AND A PHILOSOPHY

Werner Heisenberg

Abdus Salam:

In 1748 the Shahinshah of Persia, Nadir Shah, invaded India and he marched on to Delhi. He inflicted a severe defeat on the Great Mogul of India. Delhi submitted and the two kings met to negotiate peace. At the conclusion of these negotiations, which included the transfer of the famous Peacock Throne to Iran from Delhi, the Grand Vizier of the defeated Indian King, Asifjah was summoned to present to the two monarchs some wine to pledge the peace. The Vizier was faced with a real dilemma of protocol. The dilemma was this; to whom should he present the first cup of wine? If he presented it first to his own master, the insulted Persian might draw his sword and slice the Vizier's head off. If he presented it to the Persian invader first, his own master might resent it. After a moment of reflection, the Grand Vizier hit on a brilliant solution. He presented a golden tray with two cups on it to his own master and retired saying, "Sire it is not my station to present wine today. Only a King may serve a King." In this spirit I request one Grand Master of our subject, Professor Dirac, to introduce another Grand Master, Professor Werner Heisenberg.

WHEN A GOLDEN AGE STARTED

P.A.M. Dirac:

I have the best of reasons for being an admirer of Werner Heisenberg. He and I were young research students at the same time, about the same age, working on the same problem. Heisenberg succeeded where I failed. There was a large mass of spectroscopic data accumulated at that time and Heisenberg found out the proper way of handling it. In doing so he started the golden age in theoretical physics, and for a few years after that it was easy for any second rate student to do first rate work. Later I have the great fortune to do some extensive travelling with him.

In Japan, where we were very hospitably entertained I found how good he is at mountaineering and what a wonderful sense of heights he has. We had to climb a high tower with a platform at the top, surrounded by a stone balustrade. On each of the four corners the stone-work was a little bit higher. Heinsenberg climbed up on the balustrade and then on to the stone-work at one of the corners and stood there, entirely unsupported, standing on about six inches square of stone-work. Quite undisturbed by the great height, he just surveyed all the scenery around him. I couldn't help feeling anxious. If a wind had come along then it might have had a tragic result.

FIRST STEPS IN PHYSICS

W.K. Heisenberg:

I am indeed very grateful to Dirac for this very nice introduction. I would like to connect the memories I have of the old times of physics with the question about the methods which one uses in theoretical physics. There are so many attitudes which one may have: one can try to formulate phenomenological theories, one can ponder about rigorous mathematical schemes, one can ponder about philosophy and so on and I would like to analyse these various methods in connection with those experiences which I had in this life of physics.

Soon after I entered university, Sommerfeld, who at that time was the Professor of Theoretical Physics at the University of Munich, came to my room and asked me, "Well, you are interested in atomic physics, could you try to solve a problem?" I was very interested because I knew nothing about physics, but then he told me that this was quite easy and that it was more or less like solving a crossword puzzle instead of doing rigorous mathematics. The problem was this: he just had received new photographs of spectral lines in the anomalous Zeeman effect. I think that the spectra were taken by the experimental physicist Back in Tübingen, and Sommerfeld told me, "There you are, given the lines now you try, according to the theory of Bohr, to calculate or determine the energy levels belonging to these lines in order to describe every line as a difference between two energy levels and then attach quantum numbers to them and then you should be able to reproduce the picture."

One should of course try to find a formula for the energies as a function of quantum numbers and similar things. This first attempt ended at once with a catastrophe, because I found out that I had to introduce not integer numbers as quantum numbers but half integers, i.e., one half, three halves and so on, and Sommerfeld was terribly shocked to see that. He thought that it was altogether wrong and my friend Wolfgang Pauli, who was also a student in the same seminar, told me, "If you introduce half quantum numbers then you will soon introduce quarter of an integer and then a tenth of an integer and finally you will come back to the continuous analysis and we have the classical theory again." After a while there were more of us interested in these problems. There was Pauli, Hönl and others, and it turned out that one really had to introduce half integers as quantum numbers. We had a nice group of young people doing phenomenological physics together, that is, inventing formulae which seem to reproduce the experiments; in this way the formulae of Landé were found and then the multiplet formulae of Sommerfeld and Hönl and so on.

PHENOMENOLOGICAL THEORY

One of these attempts impressed me most and I think I should tell it just to show also the limitations of phenomenological theories.

Sommerfeld told me about an old paper of Voigt in Göttingen, which was written early in 1913, earlier than Bohr's Theory of the Atom. Voigt had given a theory of the anomalous Zeeman effect of the sodium D lines. For doing that he had introduced two coupled linear oscillators arranged as such that the two D lines came out, and he could also assume the coupling in such a way that he got the anomalous Zeeman effect; he could even represent the Paschen-Back effect and the intensities and quite generally he could reproduce the experiments extremely well. Sommerfeld asked me again to translate these results into the language of quantum theory, and it turned out that this was easily done. I arrived at rather complicated long formulae for the energy levels and for the intensities, formulae with long square roots containing the square of the magnetic field, the coupling constant etc., and still the experiments were represented extremely well. I have mentioned this case of a phenomenological description because it fitted so unusually well; but had it anything to do with quantum theory? Six years later we had quantum mechanics at our disposal and Jordan and I tried to calculate the same levels and intensities from quantum mechanics. We got exactly the same formulae as Voigt with the same long square roots and the same intensities. So, in one way you see that phenomenological theories can be extremely successful insofar as they can sometimes give the exact results and consequently agree extremely well with the experiments. Still, at the same time they do not give any real information about the physical content of the phenomenon, about those things which really happen inside the atom. Of course, one can finally understand it, one can say: in quantum theory, in order to calculate an anomalous Zeeman effect we have to solve a perturbation problem, represented by a "secular" determinant. Such a "secular" determinant means a set of several linear equations with several unknowns. Now two coupled oscillators are just the same thing, they also mean several linear equations with several unknowns and therefore finally one can understand that these two theories are identical in their formalism although the physical content is extremely different.

The real success of these phenomenological attempts was on slightly different line. In the course of time we tried in many cases to compare the formulae we got from the experiments with the Bohr theory. Something very funny happened. It was never possible to reproduce any of these formulae exactly by the Bohr theory. But still we got from the Bohr theory formulae, which were somehow similar to the real formulae, in the sense that, for instance, in the Bohr theory we would expect the square of the angular momentum, while empirically we had $J(J+1)$. Nowadays such a result is quite obvious because these are just representations of groups. But at that time this was a very strange result and it meant that somehow the Bohr theory was right and somehow it was quite incorrect, and we really did not know what to do about it. Because after all the quantum number, for instance, of the angular momentum was just defined as the value of the angular momentum, and it was practically impossible from classical physics that an expression like

$$\sqrt{J(J + 1)}$$

should emerge. We were quite upset about these results and at the same time we studied with extreme interest the recent papers of Bohr.

Bohr had just at that time published his papers on the periodic table of elements and we learned the very complicated structures of all the elements with ten or twenty or thirty electrons being in different orbits and we could not understand how Bohr could have obtained these results. We felt that he must have been an infinitely clever mathematician to solve such horrible problems of classical astronomy. We knew that even the problem of the three bodies had not been solved by the best astronomers, and there was now Niels Bohr who could even solve problems with 30 electrons or something like that.

BOHR'S CONJECTURE

After two years of study, in the summer of 1922, Sommerfeld asked me whether I would be willing to follow him to a meeting at which Bohr would present his theory in Göttingen. These days in Göttingen we now always refer to as the "Bohr-festival". There for the time time I learned how a man like Bohr worked on problems of atomic physics. When Bohr had given two of his lectures I dared once in a discussion to utter some criticism; I just mentioned some doubts, whether the formulae of Kramers which he had written on the blackboard could be exact; I knew from our discussions in Munich that we always get formulae which are half exact, which are partly right and partly not right so I felt that it was never too certain. Bohr was very kind and in spite of the fact that I was a very young student, he asked me for a long walk on the Hainberg near Göttingen to discuss the problem. I feel it was then that I felt I really learned what it means to work on an entirely new field in theoretical physics. The first, for me quite shocking experience was that Bohr had calculated nothing. He had just guessed his results. He knew the experimental situation in chemistry, he knew the valencies of the various atoms, and he knew that his idea of the quantization of the orbits or rather his idea of the stability of the atom to be explained by the phenomenon of quantization, fitted somehow with the experimental situation in chemistry. On this basis he simply guessed what he then gave us as his results. I asked him whether he really believed that one could derive these results by means of calculations based on classical mechanics. He said, "Well, I think that those classical pictures which I draw of the atoms are just as good as classical pictures can be," and he explained it in the following way. He said, "We are now in a new field of physics, in which we know that the old concepts probably don't work. We see that they don't work, because otherwise atoms wouldn't be stable. On the other hand when we want to speak about atoms, we must use words and these words can only be taken from the old concepts, from the old language. Therefore we are in a hopeless dilemma, we are like sailors coming to a very faraway country. They don't know the country and they see people whose language they have never heard, so they don't know how to communicate. Therefore, so far as the classical concepts work, that is, so far as we can speak about the motion of electrons, about their velocity, about their energy etc., I think that my pictures are correct or at least I hope that they are correct, but nobody knows how far such a language goes."

This was a very new way of thinking to me and it changed my whole attitude towards physics. In Sommerfeld's institute it always had seemed obvious that one should calculate something and only by rigorous calculation you can get good results.

Now coming back to the question of the phenomenological theories, I had the impression from my conversation with Bohr that one should go away from all these classical concepts, one should not speak about the orbit of an electron. In spite of the fact that you could see a track of the electron in the cloud chamber, you should not speak about the velocity or position and so on; but of course if you abandon these words then you don't know what to do. So this was a very strange dilemma and an extremely interesting situation and the question was, "What can we do in such a situation?"

After this conversation with Bohr very soon, I think half a year later, I went to Copenhagen, I worked together with Kramers on dispersion theory and again we found this funny situation, that those formulae which could be derived from the Bohr theory were almost correct, but not really correct. One gradually acquired a kind of habit how to handle such formulae, how to translate from classical physics into these phenomenological formulae. One had the impression already, that finally there must be some kind of quantum mechanics which has to replace classical mechanics. Quantum mechanics may be not too different from classical mechanics, but still it must use very different concepts.

Now in this situation it has often been said that it may be a step in the right direction to introduce into the theory only quantities that can be observed. Actually that was a very natural idea in this connection, because one saw that there were frequencies and amplitudes; and these frequencies and amplitudes could in classical theory somehow replace the orbit of the electron. A whole set of them means a Fourier series and a Fourier series describes an orbit. Therefore it was natural to think that one should use rather these sets of amplitudes and frequencies instead of the orbit.

When I came back from Copenhagen to Göttingen I decided that I should again try to do some kind of guesswork there, namely to guess the intensities in the hydrogen spectrum. The Bohr theory didn't work well for these intensities. But why should it not be possible to guess them? That was early in summer of 1925 and I failed completely. The formulae got too complicated and there was no hope to get out anything. At the same time I also felt, if the mechanical system would be simpler, then it might be possible just to do the same thing as Kramers and I had done in Copenhagen and to guess the amplitudes. Therefore I turned from the hydrogen atom to the anharmonic oscillator, which was a very simple model. Just then I became ill and went to the island of Heligoland to recover. There I had plenty of time to do my calculations. It turned out that it really was quite simple to translate classical mechanics into quantum mechanics. But I should mention one important point. It was not sufficient simply to set "let us take some frequencies and amplitudes to replace orbit quantities" and use a kind of calculation which we had already used in Copenhagen and which later turned out to be equivalent to matrix multiplication.

It was quite clear that if one only did that, then one would have a scheme which was much more open than classical theory. Of course, classical theory would be included and quantum theory also would be included, but it was much too undefined and one had to add extra conditions.

It turned out that one could replace the quantum conditions of Bohr's theory by a formula which was essentially equivalent to the sum rule of Thomas and Kuhn. By adding such a condition one all of a sudden got into a consistent scheme. One could see that this set of assumptions worked, one could see that the energy was constant and so on. I was however not able to get a neat mathematical scheme out of it. Very soon afterwards both Born and Jordan in Göttingen and Dirac in Cambridge were able to invent a perfectly closed mathematical scheme; Dirac with very ingenious new methods on q numbers and Born and Jordan with more conventional methods on matrices.

EINSTEIN ON THEORY AND OBSERVATION

I don't want to speak about details now but rather about the interpretation of the details in the sense of asking "what kind of philosophy was the most important part in this development". To begin with I thought it was probably the idea of introducing only observable quantities. But when I had to give a talk about quantum mechanics in Berlin in 1926, Einstein listened to the talk and corrected this view.

Einstein asked me to come to his flat and discuss the matters with him. The first thing he asked me was: "What was the philosophy underlying your kind of very strange theory? The theory looks quite nice, but what did you mean by only observable quantities." I told him that I did not believe any more in electronic orbits, in spite of the tracks in a cloud chamber. I felt that one should go back to those quantities which really can be observed and I also felt that this was just the kind of philosophy which he had used in relativity; because he also had abandoned absolute time and introduced only the time of the special coordinate system and so on. Well, he laughed at me and then he said, "But you must realize that it is completely wrong." I answered: "But why, is it not true that you have used this philosophy?" "Oh yes," he said, "I may have used it, but still it is nonsense!"

Einstein explained to me that it was really the other way around. He said, "Whether you can observe a thing or not depends on the theory which you use. It is the theory which decides what can be observed." His argument was like this: "Observation means that we construct some connection between a phenomenon and our realization of the phenomenon. There is something happening in the atom, the light is emitted, the light hits the photographic plate, we see the photographic plate and so on and so on. In this whole course of events between the atom and your eye and your consciousness you must assume that everything works as in the old physics. If you would change the theory concerning this sequence of events then of course the observation would be altered." So he insisted upon that it was the theory which decides about what can be observed. This remark of Einstein was very important for me later on when Bohr and I tried to discuss the interpretation of quantum theory, and I shall come to that point later.

A few words more in connection with my discussion with Einstein. Einstein had pointed out to me that it is really dangerous to say that one should only speak about observable quantities. Because every reasonable theory will, besides all things which one can immediately observe, also give the possibility of observing other things more indirectly. For instance, Mach himself had believed that the concept of the atom was only a point of convenience, a point of economy in thinking, he didn't believe in the reality of the atoms. Nowadays everybody would say that this is nonsense, that it is quite clear that the atoms really exist. I also feel that one cannot gain anything by claiming that it is only a convenience of our thinking to have the atoms — though it may be logically possible. These were the points which Einstein raised. In quantum theory it meant, for instance, that when you have quantum mechanics then you cannot only observe frequencies and amplitudes, but for instance, also probability amplitudes, probability waves and so on, and these, of course, are quite different objects.

I should also add that when one has invented a new scheme which concerns certain observable quantitives, then of course, the decisive question is: which of the old concepts can you really abandon? In the case of quantum theory it was more or less clear that you could abandon the idea of an electronic orbit.

STABILITY OF LAMINAR FLOW

But let me now leave this problem of phenomenological theories and come to the opposite question: what is the use of exact mathematical schemes? Perhaps you know, that I am not at all fond of rigorous mathematical methods, and I would like to give a few reasons for this attitude. During these years, already before the formulation of quantum mechanics, I had to do my doctor thesis. Since Sommerfeld was a good teacher, he felt that I should not always work on atomic theory; he told me, "It is not good always to walk in the mud, you should really do decent mathematical work in theoretical physics." So he suggested a hydrodynamical problem. I should calculate the stability of laminar flow. He had himself written a paper on this subject. The laminar flow between a resting and a moving wall had been treated by one of his pupils. Sommerfeld was not satisfied. This pupil, Hopf, had not been able to find a limit of stability. Experimentally, everybody knows that when the velocity gets too high then the laminar flow of a liquid goes over into a turbulent flow, eddies are created statistically, and this looks like a phenomenon of instability. Therefore it should be possible to calculate such a limit of stability. Sommerfeld suggested that I should calculate the stability of a stream of water between two fixed walls. This was my doctor thesis and I got a nice result, as I believed, namely that there was a limit of stability. At a certain Reynolds number in agreement with the experiment the flow became unstable and one got turbulent motion.

SEQUEL AFTER TWENTY YEARS

Well, I got my degree on this paper; but one year later a very good mathematician, Noether, published another paper in which he proved by very rigorous mathematical methods that this problem which I had treated had no unstable solution; the flow should be stable everywhere. That was of course a very sad result, specially considering my degree, and I always hoped that I could disprove this paper of Noether. Unfortunately I was not able to disprove it; I only had to hope for the experiments, because experimentally there was certainly a limit of stability. This problem actually took many years before it became quite clear, and I may just mention a few steps. Five years later Tollmien treated a different kind of flow and actually got a limit of stability; he could argue that his was a different problem from Noether's, so that the mathematical arguments of Noether did not apply. Then in 1944, that is 20 years after my doctor thesis, in America Dryden and collaborators made very accurate measurements of the laminar flow between two walls and the transition to turbulence. They found that really the calculations which I had made agreed well with the experiment. Lin at the M.I.T. took up the problem and confirmed the old results by new and better methods. Still some mathematicians didn't believe it, we had long discussions on this problem in 1950 at the M.I.T., and then von Neumann decided that one should use one of the electronic computers for the problem. So the biggest computer at that time finally was used for killing the problem, and it turned out that the old approximate calculations of my thesis were off the correct values by not more than about 20%, and the question was now "what about this rigorous mathematical paper". Well, the trouble is, I think, that even now nobody knows what the mistake in the paper was.

FINDING A MATHEMATICAL MISTAKE

But there was another case in which one knows where the mistake was. That was when Edward Teller came to my institute in Leipzig in 1928 or so, and he wanted to do a doctor thesis. I didn't give him a problem on turbulence, because at that time already quantum mechanics was decent physics and so I suggested that he should be interested in the H-molecule, two protons and one electron. I told him that one of Bohr's pupils, Burrau, had just published a good paper on the normal state of this molecule ion, and had found a good value for the binding energy in agreement with the experiment. Teller should try for the excited states of the molecule.

A few weeks afterwards Teller came to my room, telling me that just recently a new paper by Wilson had appeared, using very good mathematical methods, much better than Burrau's, quite rigorous mathematics, and Wilson had been able to prove that the normal state of H did not exist. Well, again this was a rather sad result and I told Teller that it must be wrong, because after all, the molecule does exist, what can we do about it. But Teller said, "Wilson's mathematics is so good, you can't say anything against it." So Teller and I, we quarrelled quite a lot about it and after, I think, about two months or so Teller actually found the error in Wilson's paper; and it was quite an interesting mistake. The error was this: The mathematical methods were actually excellent, but Wilson had argued: "We know that the Schrödinger function at far distance from the two centres must go to zero; this is correct. Therefore, our analytical function must be regular and have a zero at infinity" — which was wrong, because it was sufficient that it should go to zero on the real axis, not on the imaginary axis. Well, this is just the kind of mistake which one can make and I hope that Noether has made a similar mistake in the turbulence problem, but that I don't know.

RIGOROUS AND DIRTY MATHEMATICS

I think that you understand now why I am always a bit sceptical of rigorous mathematical methods. Perhaps I should give a more serious reason for that: When you try too much for rigorous mathematical methods, you fix your attention on those points which are not important from the physics' point and thereby you get away from the experimental situation. If you try to solve a problem by rather dirty mathematics, as I have mostly done, then you are forced always to think of the experimental situation; and whatever formulae you write down, you try to compare the formulae with reality and thereby, somehow, you get closer to reality than by looking for the rigorous methods. But this may, of course, be different for different people.

Let us now come back to quantum mechanics, and to that part of the development of a new theory, has always seemed to me the most fascinating part. When you get into such a new field, the trouble is, that with phenomenological methods you are bound always to use the old concepts; because you have no other concepts, and making theoretical connections means then applying old methods to this new situation. Therefore the decisive step is always a rather discontinuous step. You can never hope to go by small steps nearer and nearer to the real theory; at one point you are bound to jump, you must really leave the old concepts and try something new and then see whether you can swim or stand or whatever else, but in any case you can't keep the old concepts. This happened in quantum mechanics in the following way: first we had the mathematical scheme, and then, of course, we had to try to use a reasonable language in connection with it. Finally we could ask: what concepts does this mathematical scheme imply and how do we have to describe nature?

ABANDONING OLD CONCEPTS

The most difficult part in this stage of development is

abandoning some of the important old concepts. Any good physicist would be willing to acquire new concepts but even the best physicists are sometimes quite unwilling to leave some of the old and apparently safe concepts. This feeling that one cannot go away from the old concepts was also very strong in the development of quantum mechanics. You know that it has been very strong in the development of relativity, and even nowadays there are papers appearing here and there in which people just refuse to understand the theory of special relativity. They cannot understand it because they are not able to go away from the old concept of "simultaneous events". In quantum theory the same thing happened to some extent in the discussions about wave mechanics of Schrödinger and quantum mechanics. I remember one lecture of Schrödinger and the discussion afterwards in the summer of 1926. Perhaps I should mention it, certainly not to criticize Schrödinger, who was a first-class physicist, but just to show how extremely difficult it is to get away from old concepts. Schrödinger had given a lecture on wave mechanics, he had been invited by Sommerfeld and there was also Wilhelm Wien, who was an experimental physicist; at that time the theory of Bohr was not at all generally recognized as a good theory. The experimental physicists for instance in Munich disliked all this game of quantum terms and quantum jumps, they called it atomystic, that is mysticism of the atom, and they felt that it was so much unlike classical physics that it was really not to be taken seriously. Therefore Wilhelm Wien was extremely glad to hear from Schrödinger his new interpretation.

You know that Schrödinger for a time believed, that he could use wave mechanics with the same kind of concepts as Maxwell theory. He assumed that the matter waves are just three-dimensional waves in space and time like electromagnetic waves, and therefore the eigenvalue of an energy was really the eigenvalue of a vibration and not an energy. Thereby he believed that he could avoid all kinds of quantum jumps and all the rest of what he called mysticism. After Schrödinger's lecture I took part in the discussion and argued that I felt that by such an interpretation one could not even understand Planck's law. Because after all, Planck's law was based on real quantum theory, on the discontinuous changes of energy and so on. Then Wien was so angry about this remark, that he said, "Well young man, we understand that you are sorry that now quantum mechanics and quantum jumps and all the rest should be forgotten, but you will see that Schrödinger will solve all these problems very soon."

I just mentioned this episode to show how strong the feelings can be among physicists about such matters. Of course, I was completely unsuccessful in convincing either Wien or Schrödinger; but the result was that Bohr invited Schrödinger to Copenhagen. Schrödinger, in September of 1926, came to Copenhagen. Bohr, a very kind and fine man and most amiable in very way, could sometimes be almost fanatical. I remember, whenever Schrödinger stood, Bohr would also stand there, saying, "But Schrödinger you must understand, you must really." After two days Schrödinger became ill. He had to go to bed and Mrs. Bohr would bring cake and tea and so on, but Bohr would sit at the bedside: "But Schrödinger you must understand." After this time Schrödinger at least understood that it was more difficult with the interpretation of quantum theory than he had thought.

Also in Copenhagen we were not yet too happy about the interpretation, because we felt, that in the atom it seemed all right to abandon the concept of an electronic orbit. But what in a cloud chamber? In a cloud chamber you see the electron moving along the track; is this an electronic orbit or not?

QUANTUM THEORY UNDERSTOOD

Bohr and I discussed these problems many, many nights and we were frequently in a state of despair. Bohr tried more in the direction of the duality between waves and particles; I preferred to start from the mathematical formalism and to look for consistent interpretation.

Finally Bohr went to Norway to think alone about the problem and I remained in Copenhagen. Then I remembered Einstein's remark in our discussion. I remembered that Einstein had said that, "It is the theory which decides what can be observed." From there it was easy to turn around our question and not to ask, "How can I represent in quantum mechanics this orbit of an electron in a cloud chamber?" but rather to ask, "Is it not true that always only such situations occur in nature, even in a cloud chamber, which can be described by the mathematical formalism of quantum mechanics?" By turning around I had to investigate what can be described in this formalism; and then it was very easily seen, especially when one used the new mathematical discoveries of Dirac and Jordan about transformation theory, that one could not describe at the same time the exact position and the exact velocity of an electron; one had these uncertainty relations. In this way things became clear. When Bohr returned to Copenhagen, he had found an equivalent interpretation with his concept of complementarity, so finally we all agreed that now we had understood quantum theory.

EINSTEIN'S FICTITIOUS EXPERIMENTS

Again, we met a difficult situation in 1927 when Einstein and Bohr discussed this matter at the Solvay Conference. Almost every day the sequence of events was the following: We all lived in the same hotel. In the morning for breakfast Einstein would appear and tell Bohr a new fictitious experiment in which he could disprove the uncertainty relations and thereby our interpretation of quantum theory. Then Bohr, Pauli and I would be very worried, we would follow Bohr and Einstein to the meeting and would discuss this problem all day. But at night for dinner usually Bohr had solved the problem and he gave the answer to Einstein, so then we felt that everything was all right and Einstein was a bit sorry about it and said he would think about it. Next morning he would bring a new fictitious experiment, again we had to discuss it etc. This went on for quite a number of days and at the end of the conference the Copenhagen physicists had the feeling that they had won the battle and that actually Einstein could not make any real objection. I think the most splendid argument of Bohr was that he used once the theory of general relativity to disprove Einstein. Einstein had invented an experiment in which the weight of some machinery was to be determined by gravitation and so Bohr had to invoke the theory of general relativity to show that the uncertainty relations were correct. Bohr succeeded and Einstein could not raise any objection.

ELECTRONS AND THE NUCLEUS

Now I come to more recent developments. Perhaps I should, before I come to relativistic quantum theory, say a few words about nuclear physics. The only point I want to make here, again is, that it is much easier to accept new concepts than to abandon old ones. Actually, when the neutron was discovered by Chadwick in 1932, I think, it was almost trivial to say that the nucleus consists of protons and neutrons, but it was not quite so trivial to say that there are no electrons in the nucleus. The decisive point of those papers which I wrote about the structure of the nucleus was not that the nucleus consisted of protons and neutrons, but that in apparent contradiction to experiment there were no electrons in the nucleus. Everybody up to that time had assumed that there must be electrons in the nucleus, because sometimes they come out, and it was rather odd to say that they have not been in the nucleus before. Of course, the idea was that the short range forces between neutron and proton somehow might have to do with the creation of electrons in the nucleus. Anyway it seemed to me a good approximation to assume that such light particles cannot exist in the nucleus. I remember that I have been criticized very strongly for this assumption by extremely good physicists. I got one letter saying that it was really a scandal to assume that there were no electrons in the nucleus because one could just see them coming out; I would bring a complete disorder into physics by such unreasonable assumptions and they could not understand my attitude. I just mentioned this small event, because it is really difficult to go away from something which seems so natural and so obvious that everybody had always accepted it. I think the greatest effort in the developments of theoretical physics is always necessary at those points where one has to abandon old concepts.

CHANGING THE OUTLOOK OF ATOMIC PHYSICS

May I now turn to the problem of the elementary particles. I think that really the most decisive discovery in connection with the properties or the nature of elementary particles was the discovery of antimatter by Dirac. That was an entirely new feature which apparently had to do with relativity, with the replacement of the Galilei group by the Lorentz group. I believe that this discovery of particles and antiparticles by Dirac has changed our whole outlook on atomic physics completely. I do not know whether this change was realized at once at that time, probably it has been accepted only gradually; but I would like to explain why I consider it as so fundamental.

We know from quantum theory, that for instance a hydrogen molecule may consist of two hydrogen atoms or of one positive hydrogen ion and one negative hydrogen ion. Generally one can say that every state consists virtually of all possible configurations by which you can realize the same kind of symmetry. Now as soon as one knows, that one can create pairs according to Dirac's theory, then one has to consider an elementary particle as a compound system; because virtually it could be this particle plus a pair or this particle plus two pairs and so on, and so all of a sudden the whole idea of an elementary particle has changed. Up to that time I think every physicist had thought of the elementary particles along the line of the philosophy of Democritus, namely by considering these elementary particles as unchangeable units which are just given in nature and are just always the same thing, they never change, they never can be transmuted into anything else. They are not dynamical systems, they just exist in themselves.

After Dirac's discovery everything looked different, because now one could ask, why should a proton be only a proton, why should a proton not sometimes be a proton plus a pair of electron and positron and so on. This new aspect of the elementary particle being a compound system has at once looked to me as a great challenge. When later I worked together with Pauli on quantum electrodynamics, I always kept this problem in my mind.

PAIR CREATION

The next step in this direction was the idea of multiple production of particles. If two particles collide then pairs can be created; then there is no reason why there should only be one pair; why should there not be two pairs. If only the energy is high enough one could eventually have any number of particles created by such an event, if the coupling is strong enough. Thereby the whole problem of dividing matter had come into a different light. So far one had believed that there are just two alternatives. Either you can divide matter again and again into smaller and smaller bits or you cannot divide matter up to infinity and then you come to smallest particles. Now all of a sudden we saw a third possibility: we can divide matter again and again but we never get to smaller particles because we just create particles by energy, by kinetic energy, and since we have pair creation this can go on forever. So it was a natural but paradoxical concept to think of the elementary particle as a compound system of elementary particles. Of course then the problem arose, "What kind of mathematical scheme can describe such a situation?"

At that time one knew from Dirac's theory of radiation and from the attempts of Pauli, Weisskopf and myself that one had great difficulties in avoiding infinities in quantum electrodynamics and, more generally, in quantum field theory with interaction. I agree completely with Dirac in disliking infinities in the sense that if you introduce infinity in physics, you just talk nonsense, it cannot be done. Therefore I tried to think of mathematical schemes in which you can avoid infinities from the very beginning. Again, I remembered the old story of the observable quantities and therefore I felt that it was probably useful to ask "what can we really observe in a collision between elementary particles", and so it was natural to come to the S-matrix and to say that the S-matrix or scattering matrix is a rational basis for a theory.

Again of course it is considerably easier to go this first step, namely to say that such and such things can be observed, than to go the next step and to narrow down the assumptions. But finally you have to make new assumptions and end up by saying "such and such things cannot be observed anymore". So the question was now, "How can we narrow down the concept of the S-matrix in order to get something which is really workable, in which we can define what we mean, in which we can formulate natural laws." Well, at that time I had learned again from Dirac that one could perhaps work with a field theory making use of an indefinite metric in Hilbert space. I knew of course that Pauli had criticized that very strongly, as he always did, — and very often successfully — namely by saying that if one had an indefinite metric in Hilbert space it meant negative probabilities, and therefore such a scheme would not work.

It was natural to think of the possibility, that for the asympototic region of course you must have positive probabilities, therefore asymptotically you must have a unitary S-matrix; but at the same time it would be allowed to go away locally from this concept of probability and say, "Locally we may have negative probabilities because locally we cannot measure anything in the same way as in the asymptotic region." The concept of probability may fail when we get below a certain "universal length". Therefore I tried to narrow the scheme down by saying, "There shall be local field operators but these operators may work in a Hilbert space which has not an ordinary metric but may have an indefinite metric." The advantage of this scheme was that one could actually avoid infinities but, of course, at a very high price, namely at the price of losing the definite metric in Hilbert space. On the other hand by that time the whole scheme already looked rather convincing to me, because the experiments in the meantime had proved that there was actually multiple production of particles.

PHENOMENON ESTABLISHED

That had in fact been a rather controversial subject for more than ten years, because there were, of course, these cosmic-ray showers which everybody had known since 1936 or so. But these showers could be very well explained by the cascade theory of Bhabha and Heiter; so there was no proof for multiple production of particles. It was not before about 1950 that one really could get very good evidence for the existence of this multiple production. But since this phenomenon was now well established, I felt that one could go on in the same direction and therefore I tried to formulate a kind of field theory. I thought that for the mathematical scheme the model of Lee could give some help, but of course I was very well aware of the fact that in field theory we had no rigorous mathematical scheme. I felt that it might be sufficient for the time being to look for a mathematical scheme which somehow fitted the experimental situation.

To begin with we did not know of any good field equation which could represent the actual situation and the experiments. But then in 1957, after a lecture I had given at Cern, I met Pauli and we discussed the new possibilities. We had learned from Lee about the non-conservation of parity in β-decay, and following up this line of ideas we came to a field equation which turned out to contain the SU(2) group, namely the isospin. Pauli was more enthusiastic about this possibility than I have ever seen him. I got letters from him saying that now a new dawn of physics has begun and all our difficulties will disappear very soon and so on. I always had to stop him and say, "Well, that is not so easy." But he was so excited about it and full of energy and enthusiasm, his central interest was to work on these problems.

During this period I met him several times in Zurich, but then he had to go to the States. When he had to give lectures on these problems there he tried to rationalize his feelings and then he felt that he was not able to do it; he saw that the whole problem was much more complicated than he had hoped for. I should perhaps mention in this connection that the most essential idea that Pauli had contributed to our common paper, was (in a somewhat preliminary form) the idea of a degeneracy of the ground state which later, in connection with the theorem of Goldstone has played a considerable role in elementary particle physics.

PAULI'S CRITICAL ACUMEN

Pauli's whole character was different from mine. He was much more critical and he tried to do two things at once while I would think that this is really too difficult even for the best physicist. He tried first of all to be inspired by the experiments and to see in a kind of intuitive way how things are connected, and at the same time he tried to rationalize his intuitions and to find a rigorous mathematical scheme so that he really could prove everything what he said. Now this is, I think, simply too much, and therefore Pauli has through his whole life published much less than he could have published if he had abandoned one of these two postulates. Bohr had dared publish papers which he could not prove and which were right after all. Others have done a lot by rational methods and by good mathematics, but the two things together that I think is too much for one man. Pauli was completely disappointed, when he saw the difficulties, and so he gave

up in a rather sad way. He told me that he felt that his thinking was
not strong enough any more and that he was not well at all; but he
encouraged me even after he had withdrawn his approval for the
publication. I should go on, he said, but he could not continue and as
you know, unfortunately he had to die half a year later. This was
rather a sad end of my long friendship with Pauli and I can only say
that I regret even now almost every day that I cannot have his very
strong criticism which has helped so many times in my life in physics.

But let us come back now to the further developments in
physics. I think we know more about the degeneracy of the ground
state and perhaps most of you will know much more of the details and
of the mathematical scheme than I do. I can only hope that the picture
will remain a closed picture. I cannot doubt that it is possible to
describe the whole spectrum of the elementary particles in the same
way as we describe the spectrum of say, the iron atom in quantum
mechanics, that is by one unified natural law. This law of course will
be a kind of summary, a résumé of the many details which are being
studied now.

MY GENERAL PHILOSOPHY

It may be tempting to add a prescription about how we should
work in theoretical physics. This would however be very dangerous,
because the prescription ought to be different for different physicists.
Therefore I can only speak about that prescription which I have
always used for myself. This was that one should not stick too much to
one special group of experiments; one should rather try to keep in
touch with all the developments in all the relevant experiments so that
one should always have the whole picture in mind before one tries to
fix a theory in mathematical or other languages.

I might perhaps describe this general philosophy by telling two
quite different stories. When I was a boy, my grandfather, who was a
handicraftsman and knew how to do practical things, once met me

when I put a cover on a wooden box with books or so. He saw that I took the cover and I took a nail and I tried to hammer this one nail down to the bottom. "Oh," he said, "that is quite wrong what you do there, nobody can do it that way and it is a scandal to look at." I did not know what the scandal was, but then he said, "I will show you how you could do it." He took the cover and he took one nail, put it just a little bit through the cover into the box, and then the next nail a little bit, the third nail a little bit, and so on until all the nails were there. Only when everything was clear, when one could see that all the nails would fit, then he would start to put the nails really into the box. So, I think this is a good description of how one should proceed in theoretical physics.

The other story concerns the discussions which Dirac and I had. Dirac often liked to say — and I always felt that it was a slight criticism, — he felt, that one can only solve one difficulty at a time. This may be right, but it was not the way I looked at the problems. Then I remembered that Niels Bohr used to say, "If you have a correct statement, then the opposite of a correct statement is of course an incorrect statement, a wrong statement. But when you have a deep truth, then the opposite of a deep truth may again be a deep truth." Therefore I feel that it is perhaps not only a deep truth to say, "You can only solve one difficulty at a time," but it may also be a deep truth to say, "You can never solve only one difficulty at a time, you have to solve always quite a lot of difficulties at the same time," and with this remark perhaps I should close my report.

THE SCIENTIST AND SOCIETY

Eugene Paul Wigner

Professor Henry Smythe, Ambassador of the USA to the IAEA in Vienna and Professor of Physics at Princeton, referred to by Wigner as an excellent administrator of science and as a man of unusual absolute integrity, introduced Professor Wigner. He recalled that when Wigner first came to Princeton with John von Neumann, they had felt for some time that they had a pretty good experimental group, but did not think they had much of a good theoretical physics group there. With their coming, the situation changed immediately, and had continued to change. Professor Wigner had contributed enormously to the advances of theoretical physics. He had also contributed greatly to the strength of the Department of Physics at Princeton and particularly of the theoretical side.

FIRST SOURCES

E. P. Wigner:

I wish to tell you about what I believe is the fundamental motivation of the scientist. How the life of the scientist has changed during the period I have tried to be one, what he can expect from and what he should do for the society which enables him to enjoy a life of physics.

First, I want to mention a few sources from which I learned a great deal that is relevant to my present subject. My first teacher was Polanyi but if I enumerated everything that I learned from him I would never get any further. The next source was Wilhelm Ostwald's *Grosse Männer*. This is a collection of stories of several great scientists, with introductory remarks of a general nature giving a distillate of universal verities which he obtained when studying the stories of his heroes. Next in the line of my recollections are three long walks with James Franck, in Princeton, during the early but already gloomy days of the Hitler regime. We discussed just those questions about which I wish to talk today. Last there have been some recent conversations with historians and philosophers of science, including Dr. Mehra who is among us now, which helped me greatly to clarify my views.

REGULARITIES IN A COMPLICATED WORLD

A friend of mine likes to quote me as saying that what I wish to accomplish in life is to leave a bit more order and understanding behind than I have found. I do not remember when I told him that but there is a great deal of truth in it. We have a complicated world around us, full of unforeseeable events, and it is calming to the soul to find and know something that is orderly and unchangeable. This is not all. If we think a little further about our relations to the world, we soon realise that if we could not find regularities in it, we could not influence the events around us. The regularities in question are connections between subsequent events such as that this eraser in my hand will fall down on the table if I let go of it above the table. If there were no such regularities, we could not exert influence on the events — I could not produce a thumping noise with the eraser and I could not see it jump if I did not know that it will have this effect when I let go of it. Hence, the regularities make life possible in the sense I believe we understand life, to have an influence on events.

Of course, the regularities in which we physicists are interested are much more subtle. Nevertheless, I believe that the basis of the motivation, and of the striving to recognise some order, is common to all living beings and that, in fact, it is closely connected with the essence of life.

The question then presents itself, "What are the limits of our search for regularities?" Would we be happiest if the regularity were complete so that we could foresee everything and know and understand everything? If the preceding analysis of the causes for our search is correct, the answer must be negative. If the order were complete, if we could foresee everything, we would be again in the situation in which we could not influence anything, in which all would be determined and our will and our desires would have no way to manifest themselves. Hence, in this sense, the existing world is the best one: there are some regularities, and we need them for what we call life. But there are plenty of irregularities, and they are equally indispensable for what we call life.

"UNREASONABLE ACCURACY" OF PHYSICAL LAWS

This situation is magnificently reflected in physics. We have initial conditions which show no regularities, and there are laws of nature which express miraculously precise regularities. There is, however, a much sharper distinction between the domain of regularities and of arbitrariness than we had any reason to expect and this is, perhaps, the most remarkable result of physical theories. Charles Pierce, the philosopher, commented on the unreasonable accuracy of physical laws and now Dr. Dirac has re-emphasized the fact that, offhand, we had no reason, and no indication, to expect laws of physics to be as accurate and, in a sense, as simple, as we have found them to be. Thus, in a deeper sense, science, far from having abolished miracles, has recognised and drawn attention to a miracle of overwhelming power which holds us scientists in awe and in bondage. More so, much more so, than people in other professions.

The desire for an order manifests itself not only in our striving to recognise regularities in nature, in the succession of events, but also in the structures which we have ourselves created, our theories and concepts. Mathematics is entirely devoted to the search for regularities in the relations between concepts, created just for this purpose. But physical theories also have an intricate structure, and the elucidation of this structure, for instance, the recognition of the parts of the theory which are responsible for a certain conclusion, is also providing us with a great deal of satisfaction. The discovery of Klein and Noether, that a conservation law for energy is valid in any mechanics with a time independent Lagrangian (a theory of motion) must have given them a feeling of elation, the feeling of being suddenly cleaner and freshly bathed. Those of us who were not favored by a discovery such as that of the elementary electric charge and the existence of its carriers in metals, or of the equation that most adequately describes this carrier, can derive, and have derived, enough satisfaction for a lifetime by having clarified, perhaps not the structure of the events, but at least the structure of the theories which are a condensation of the regularities between the events. The pleasure experienced in this way has much in common with the pleasure of the mathematician. It is a real pleasure, nevertheless.

CONSEQUENCES OF THE SCIENTIST'S WORK

The scientist's activities satisfy not his desire to influence the world around him, but a sublimation, an ideal of this desire. I believe this is true. Nevertheless, it is clear that, surprisingly frequently, he does influence the world around him. Without modern science, we would have no radios, no television, no automobiles for the students to makes barricades of, no antiballistic missiles. These are very real consequences of the scientist's activities. Nevertheless, I do hold on to what I said because effects are consequences of, rather than motives for, the scientist's activities. In fact, some of our fellow scientists are unhappy when they learn that their results and conclusions have been used to produce a new drug, or some new equipment. They feel that their sublimated desires are somehow condensed and they feel that their pure and sublime science has been debased by being applied to the benefit of the society which should support them without reaping such benefits. I do not agree with this attitude, but surely it proves that the pure scientist's motive is a sublimation of the instinctive desire to influence the course of events, and not the desire itself.

Are there some negative traits in the makeup of the scientist which make it easier for him to turn away from the goals which most of his fellow citizens pursue, to refuse the participation in the quests which inspire most of his friends and acquaintances?

It seems to me, but I am less than certain in this, that his desire for influence is sublimated to such an extent that the common, everyday desire for power and influence is smaller than it is on the average. Until a few years ago, I believe that few of us thought much about the unfortunately very widely spread craving for power and influence. When the frequency of this craving dawned upon me, about six or seven years ago, I brought up the subject with colleagues, and with friends outside the world of physics. Most of my colleagues did not understand what I was speaking about, and most of my non-physicist friends did not understand why I was talking about the matter, it was such an obvious fact to them. I then recalled many observations on the subject which I had heard in the past, including my father's explanation of the reason for so many people's coveting of great wealth, and a number of events which had been mysterious to me became clearer.

At any rate, I believe that at least the scientists who are my contemporaries had a good deal of inclination towards retirement from the struggles which go on in our society, a certain fondness toward the monastic way of life and that indeed this was characteristic for those who chose Science as their vocation. Franck said, on one of our joint walks, that we scientists use science as an opiate enabling us to forget what goes on around us and to disclaim responsibility therefore. The young scientists of those days wanted to learn in seclusion, create new ideas in solitude and retirement.

CHANGES IN SCIENCE

Whether this characteristic of withdrawal and a penchant for a monastic life, is true of the present day scientist to the same extent as of the scientist thirty or more years ago, is not sure. This brings me to the next subject, the great changes that took place in science during my own life of physics.

I believe I was 17 when my father asked me what I thought I would do with my life. I expressed the desire to become a scientist, a physicist by preference. He must have suspected that and, at any rate, his answer was, "Hm, how many positions for physicists are there in the whole of Hungary?" I gave a somewhat exaggerated figure and said, "Four." He overlooked my exaggeration and asked me whether I expected to get one of those four positions. We agreed that it might be best if I studied something of greater practical value such as chemical engineering and, indeed, this is the subject in which I acquired a degree. However, during the relatively short period which elapsed between my 17th year and the granting of my degree, the world changed a great deal. First, it shrank, the distance between Germany and Hungary decreased, not so much in travelling time as spiritually, and the idea to assume a position outside of Hungary did not appear so absurd any more. Second, the number of positions for physicists increased greatly. Polanyi, my doctor-father, had a serious conversation with my father and myself pointing out that a career in science did not appear something romantic any more and that we should seriously consider it. Indeed, the status of the scientist had changed enormously during the six years in question. In 1919 he was regarded, at least in Hungary, as a venerable but very queer bird. By 1924 it became a career which did mean a great deal of retirement from the world but nevertheless a career which could be seriously considered in Germany. Even in Hungary the smile it evoked became a smile of tolerance.

This development has continued ever since. Maybe I am old-fashioned when I expect people choosing a career in science to do this without the expectation of obvious outside rewards, in the spirit of craving for a life of learning and, hopefully, creativity. The fact is that many of our young men choose the scientific career in just this spirit but the fact is also that many others expect outside rewards, influential positions, high distinctions and a life of what we call success. I do not know the spirit of which group will ultimately prevail. Perhaps there will be a mixture of the two, perhaps those in the more self-asserting group will eventually leave science and assume administrative positions inside, or outside, academic life. But surely the spirit and the traits which were taken for granted in a scientist earlier in this century cannot be taken for granted any more — the scientist of today is, in his attitude toward life, more similar to his non-scientist contemporary than was the scientist of thirty years ago. This is neither necessarily good, nor necessarily bad, it may be even less of a change than it seems to me, but there surely is some change. The self-assurance of today's physicist is very different from the attitude his older colleague exhibited in his youth — he was almost apologetic for the unconventionality of his interests and strivings.

EMERGENCE OF BIG SCIENCE

Another very significant change is the emergence of big science, that is laboratories with several thousand members. We all feel that being a scientist in such a laboratory is very, very different from being a scientist who works in solitude, that the use of a 70 BeV accelerator by a team of a score of scientists, approved by the administrative

committee of the accelerator, is very different from the contemplative
life that was the essence, though perhaps not the whole, of science as
late as the early part of this century. I do not want to discuss what
Alvin Weinberg has called big science in detail. It is clear that it has
accelerated the acquisition of knowledge enormously. It is also clear
that it needed the less retiring scientist, with the more conventional
and more aggressive attitude which I described.

Having spoken about the years preceding my becoming a
physicist, it would be the right thing to continue and to tell you about
my development and the work on which gave me most pleasure.
However, it would be difficult to review my work. Somebody said that
I have made infinitesimal contributions to an infinity of subjects. This
is, of course, an unjust accusation; I have not contributed to infinitely
many subjects.

COURAGE IN GUESSING

My doctoral dissertation was an attempt — which has turned
out later to have been correct — to calculate the rate of chemical
association reactions, such as the one mentioned in this symposium by
Dr. Salpeter: two hydrogen atoms colliding and forming a molecule.
There were two problems. If we consider the collision of the atoms in
the center of mass-coordinate system, the two atoms have to form a
molecule at rest and the energy of the molecule is quantized. It is then
infinitely unlikely that the kinetic energy of the atoms be just so large
that the energy of the system coincides with one of the energy levels of
the molecule. Born and Franck, in a joint paper, also made this point.
It was concluded, therefore, that the association reaction was infinitely
unlikely. The situation was worse than this: the angular momentum of
the molecule was also quantized and it was equally unlikely that the
atoms which collide have just the right amount of angular momentum
about their center of mass. All this was, of course, years before
quantum mechanics was discovered. It would have been, therefore,

natural to conclude that simple association reactions are impossible, or have zero probability, had there not been a wealth of experimental information available, from actual chemical reactions, that they actually do take place. The solution to the problem which I proposed, on the basis of experimental information and the study of the establishment of the chemical equilibrium was then (a) that the energy levels are not sharp but have a certain breadth and that the reaction can take place if the energy of the colliding pair of atoms falls within that breadth and (b) that the limitation with respect to angular momentum should be disregarded, the angular momentum of the pair being filled up automatically and mysteriously to the next integer multiple of Planck's quantum $\hbar$. These two prescriptions then guaranteed the proper establishment of the chemical equilibrium of dissociation. They also give a fair picture of resonance reactions in general and I remained interested in these reactions, as most of you know. I told this story because I thought you might be interested in some other corner of the picture of the frame of mind of people in the pre-quantum mechanical days. One had to guess more at that time than demonstrate and one's courage in guessing was much greater than it is now when the inadequacy of the available theory is not established. Nothing that I said contradicts, of course, Dr. Salpeter's conclusion that the simple association reaction, the forming of an H molecule from the collision of two H atoms, is a very, very unlikely process: the energy levels of the H molecule are narrow and far from each other. I calculated the rate of the formation of the molecule as a result of a collision of three H atoms very much later.

THE PLEASURE OF EXPLORATION

Having heard the story of one of my calculations, I am sure you do not want to hear the story of all the others. I really cannot tell which gave me most pleasure. I always enjoyed the work and when and if I was able to conclude it, I always felt that there was a bit more order in my mind and thinking. The same was true, more often than not, after reading an article which I could understand and in many cases I felt afterwards a high elation, almost a euphoria. Furthermore, the pleasure of exploration has not diminished in the many years that I have enjoyed it. Age brings a happiness and relaxation and as long as one is not constantly reminded of one's failing powers, is the happiest period of life. Let me just add that, except for the concern for the success of the work, and the deep concern about the eventual outcome of the war, the work for the government during the war was also interesting and satisfying. The friendships I formed as a result of association with other physicists is also a continuing source of pleasure and satisfaction.

The last subject on which I wanted to share some thoughts with you, is the relation between scientist and society.

As long as there were four physicists in a population of seven million, this relation was not of major importance. Now, however, when the U.S., for instance, spends 20 billion dollars a year on research, out of a total national income of 800 billion dollars, so that directly or indirectly, about 5 million people work on research of one kind or another in a country of 200 million, the importance of the question has a different order of magnitude. This remains true even if you find one reason or another — and there are such reasons — to change any numbers by a considerable factor.

THE PRIVILEGE OF A SATISFYING LIFE

What I am advocating is that we realise how much we owe to society. It keeps us — and if I look around myself I find that it keeps us in luxury — for doing what we want to do anyway, for doing what gives us most pleasure. I believe that we should show, in return, some helpfulness and be less than annoyed if one of our conclusions or discoveries finds a practical application. The book of Ostwald points out that almost every one of his Great Men has, at one time or another, devoted time to some practical problem, to the combatting of disease, the increase of production, or something similar. He also points out that almost every one has devoted, usually toward the end of his career, time to advise his government on questions of the administration of a scientific enterprise, and on the possibility of practical applications thereof. We, who are generously supported by our society, should show a sense of humility and gratitude rather than contempt for the non-scientist. I know that it can be argued that society derives benefits from supporting us — but so does the man who jumps into the water for rescuing another. I find, therefore, statements of the sort "the worth of the society can be well judged by the extent to which it supports its scientists adequately" simply repelling. Such statements naturally provoke counterstatements like that of Professor Harry S. Johnson's. He said: "The argument that individuals with a talent for research should be supported by society differs little from arguments formerly advanced in support of the rights of the owners of landed property to a leisured existence, and is accompanied by a similar assumption of superior social worth of the privileged individual over the common men." I believe we should do all we can to avoid such criticism; the resulting confrontation can do only harm, harm to both society and to science, particularly big science.

FROM MY LIFE OF PHYSICS

Oscar Klein

Professor V. Fock, Physical Institute of the University of Leningrad, USSR:

Professor Klein's activity in the field of physics began about the same time as mine — in the twenties — or even earlier. From about 1920 he worked for a long time with Niels Bohr. Professor Klein's name met with mine in 1926, when we published independently and nearly at the same time (Klein somewhat earlier) an attempt to find a relativistic wave equation for a charged particle which would generalize de Broglie's and Schrödinger's wave equations. The spin was then not taken into account and the true equation for the electron was found later by Dirac, but the equation that bears (as it undoubtedly should) Klein's name can be interpreted as the equation for particles with zero spin (for bosons) and is still of importance.

Klein's name is also associated with the relativistic formula for the scattering of electromagnetic radiation by free electrons (the Klein-Nishina formula).

Much discussed in the early thirties was the Klein paradox connected with states of negative energy. The solution of this paradox came with Dirac's theory of positrons.

Klein's transformation of the commutation relations and his ideas on the isotopic spin were many years later developed by Yang-Mills. Professor Klein's scientific interests in Einstein's relativity and gravitation theory never ceased. In his early work he investigated questions connected with the five-dimensional formulation. His general attitude towards unlimited cosmological applications of Einstein's gravitation theory is, so far as I know, critical.

Last I must mention another important part of Professor Klein's work, namely his activity on the Nobel Prize Committee.

FROM CHILDISH CURIOSITY

O. Klein:

Being received with so much kindness and generosity at this coast of the Adriatic sea brings to mind the arrival of Odysseus at a land — situated in this same sea — after much toil and suffering. It lies very near to my heart to think of this, having loved and often reread the Odyssey ever since my childhood — in Swedish translation. And some years ago I was given a Greek edition with an English translation, from which I can guess at the meaning of some of the Greek words. As you know, Homer stresses over and over again the tribulations through which his hero had to pass. Now, none of us physicists has conquered Troy, but some have solved problems requiring a similar kind of ingenuity as that used by the man of many devices. A main trouble for us theoreticians rather resembles that represented by Charybdis and Scylla, between which Odysseus was forced to steer. Thus, speculation is certainly a necessary part of theoretical work, just as much as building on experimental facts. Still, it drags many of us into a mental whirlpool not unlike the hydrodynamical one of Charybdis, from which the escape feels like a miracle. On the other hand, sticking too closely to facts — Scylla had six hard ones — may be equally deadly, when using them as building stones for a theory. There are many examples of this even from the days of Aristotle.

Returning to the reception at this friendly coast, you will remember that in Odysseus' time a stranger would be met with a number of questions like: from where do you come, what is your name, who are your parents, what sort of life do you lead, a pirate's or a peaceful one? Now, is this not almost the evening speakers' situation? I must try to make the best of it, my answer having one advantage over that of Odysseus, which lasted many hours, filling about a quarter of the Odyssey, namely comparative shortness.

My name and whence I come are already answered. To the question about parents I would have liked to give a much fuller answer than time will allow, because of their deep influence on my whole outlook on life not to be separated from that on physics. Some eleven years before I was born my parents moved from Germany to Stockholm, where my father became chief rabbi at the Jewish congregation. He was born in the little town of Humennéh in the Carpathian mountains, his parents having a small shop on its outskirts. He left home at an early age, studied the Talmud in Eisenstadt and got his doctor's degree in Heidelberg, finishing his rabbinic studies under the guidance of Abraham Geiger, the famous founder of a deeply liberal movement in Judaism. There was no dividing line between my father's way of regarding his profession and his deep engagement in research, mostly concerning the origin of Christendom in relation to contemporary trends in Judaism. Being very far from a narrowly literal belief in the Bible text, he shared the dream of the Hebrew prophets of an era of universal peace for mankind. He died shortly before the First World War, which spared him the deep disappointment it caused to others of a similar mind.

In contrast to many theoretical physicists, my life in physics had not a philosophical background but came as a result of childish curiosity. On the one hand this made me for several years a bit of a young naturalist, collecting all sorts of things from shells and butterflies to stars — the latter with the help of my mother's opera glasses. On the other hand it made me ask my seven-year older brother — a great authority for me — all sorts of impossible questions about connections and origins, somewhat like Kipling's elephant child.

WHEN ALL WAS NEW

By and by I began to read scientific books, at first mostly on biology, then on chemistry, my father helping me to find such books, among them Darwin's *Origin of Species* and *Descent of Man*. Just before I was 16 (summer 1910) he introduced me to the great physical Svante Arrhenius, whom he knew "als ein humaner Mensch" a high praise in his mouth. Arrhenius kindly let me do some experimental work in his laboratory, lent me books and advised me as to reading. These were wonderful times when all was new and my eagerness almost unlimited.

In my last year at Arrhenius' laboratory (1917-18), after I had finished my university studies and also my rather long military service, I spent much time reading papers on the new quantum theory and also on statistical mechanics. Thereby I was, of course, amazed by Bohr's explanation of the spectroscopic Rydberg constant by means of a simple atomic model. But I was far from understanding the deep background of this result, being more impressed by the explicitly mathematical papers of Sommerfeld, Einstein and Debye. So, when I was granted a fellowship for studies abroad, I chose primarily Einstein and Debye, the latter more for dipoles, on which I had been working, than for quantum theory. But, since Bohr was so near, I wrote to him first, asking whether I might come for a shorter stay in Copenhagen, which he kindly accepted. This was in the spring of 1918, and there I went and stayed, with some interruptions, most of them in Stockholm, the longest at the University of Michigan, until I started as a professor of theoretical physics at the University of Stockholm in the winter of 1931. And I never came to Einstein and Debye.

When I came to Copenhagen, Bohr had only one, but a very efficient, theoretical collaborator, namely Kramers, and he had no institute but just a room at the Engineering College, and we were allowed to sit in the library there. The first part of his paper on spectral lines had just appeared, which, with Kramers' help, I studied thoroughly. And Bohr himself commented further on his ideas, especially during a long walking trip north of Copenhagen, when he also talked about many other things more or less connected with physics: his general philosophical attitude and his father's view on the relation between life and physics, both themes containing the germs of his later complementarity viewpoint.

I was then toiling with the forces between ions in strong electrolytes on the lines of Bjerrum and dipoles on the line of Debye, trying to apply Gibbsian statistical mechanics to these problems; and Bohr showed me his deeper view of this subject, telling me how Gibbs' general canonical distribution gave the very definition of temperature. All this meant a new epoch for me, and an essentially happy one, although I had more troubles than results from my own work, which, however, led to my thesis on generalized Brownian motion, meant as a foundation for a theory of solutions of interacting particles.

During these early years the essential work of Bohr and Kramers turned around quantization, the way of sorting out, among the mechanically possible orbits of electrons in atoms, those suitable to represent the quantum states — although Bohr already knew that this could be but a provisional procedure, a preparation for a rigorous quantum mechanics. But astonishingly much was reached in this way — for which Bohr introduced the word correspondence — through his exceptional gift of obtaining largely correct results from unfinished theory and insufficient experimental facts, verily a strangely successful passage between Charybdis and Scylla.

DEEPER BACKGROUND

In these surroundings it was natural to dream about the deeper background to these strange quantum rules with their whole quantum numbers. I shall dwell a little on this because it was the primary source of several attempts of mine, which slowly, after many mistakes, led me to my present views on relativity theory and its relations to quantum theory and cosmology.

Some time in the early twenties, from studying Fresnel's work on the wave theory of light, I was struck by the fact that whole numbers in physics appear in one of two ways, either through atomism, the usual way of regarding quanta, or through wave interference. Slowly this trend of thought began to take a somewhat more definite form due to my reading the review in Whittaker's *Analytical Dynamics* of Hamilton's original way to what was later called the Hamilton-Jacobi equation—known to us as quantizers from Jacobi's *Vorlesungen über Mechanik*. According to Hamilton, who started from optics, this equation determines the propagation of a wave front in the manner of Huygens, forming a link between geometrical optics and mechanics. Thus a quantum state belonging to a particle moving in a closed orbit came to look as if a wave (in the geometrical approximation) interfered with itself, the number of wavelengths on the orbit being equal to the quantum number. I had known this for some time, when in the summer of 1923, in connection with my writing a little book on optics, I noticed that this could be interpreted as the condition for a standing wave, i.e., a proper vibration.

Towards the end of that summer I got married and in the beginning of September my wife and I went to Ann Arbor in the U.S.A., where through the mediation of Walter Colby — now for many years a dear friend of ours — I had been appointed an instructor in theoretical physics at the University of Michigan. There we stayed for almost two years during which time, apart from lectures and other teaching, I worked hard on different problems, more or less connected with the study of molecular spectra going on at the Department of Physics under the most competent and unselfish guidance of H.M.

Randall, the head of the department.

WHIRLPOOL OF SPECULATION

The next autumn, however, I gave a lecture course on electromagnetism, towards the end of which I derived the general relativistic Hamilton-Jacobi equation for an electric particle moving in a combined gravitational and electromagnetic field. Thereby, the similarity struck me between the ways the electromagnetic potentials and the Einstein gravitational potentials enter into this equation, the electric charge in appropriate units — appearing as the analogue to a fourth momentum component, the whole looking like a wave front equation in a space of four dimensions. This led me into a whirlpool of speculation, from which I did not detach myself for several years and which still has a certain attraction for me.

The strong impression this made on me came from the attempt I mentioned to find a wave background to the quantization rules. Thus, for some time I had played with the idea that waves representing the motion of a free particle had to be propagated with a constant velocity, in analogy to light waves — but in a space of four dimensions — so that the motion we observe is a projection on our ordinary three-dimensional space of what is really taking place in four-dimensional space. This idea seemed to me to get certain support by some occasional utterings of Bohr. Thus, while most strongly stressing the impossibility of describing quantum phenomena by a space-time theory of the ordinary type — regarding them as "a knot on existence which might be shifted but not removed" — he sometimes said that such a theory would perhaps be possible in a space of four dimensions. This had nothing to do with so-called parapsychological phenomena, of which, since my early youth, I have remained most sceptical — like Bohr.

Apart from being entangled in these speculations there was another reason which made me hesitate to start an investigation, by means of a wave equation, of the standing waves corresponding to the quantum states of a particle in a field of force. Thus, from the book of Hadamard on wave propagation I knew that the relation between a wave-front equation and a second-order wave equation is not unique, that on top of the linear second-order terms there might be non-linear terms. And I believed that such terms might have some bearing on the particle nature of electrons. This is certainly foolish, but was not obviously so at that time. But Dirac may well say that my main trouble came from trying to solve too many problems at a time!

I did not try until more than a year later, when again in Copenhagen, to determine the states of standing waves of the linear wave equation corresponding to the Hamilton-Jacobi equation of a harmonic oscillator. But due to lack of time and poor mathematics I had not succeeded in this when Schrödinger's paper containing the similar solution for the electron in the hydrogen atom appeared.

IN FIVE DIMENSIONS

Returning to my Ann Arbor attempts, I became immediately very eager to see how far the mentioned analogy reached, first trying to find out whether the Maxwell equations for the electromagnetic field together with Einstein's gravitational equations would fit into a formalism of five-dimensional Riemann geometry (corresponding to four dimensions plus time) like the four-dimensional formalism of Einstein, which, by the way, I knew very superficially at the time and now tried to learn by means of Pauli's excellent book. It did not take me a long time to prove this in the linear approximation, assuming a five-equation, according to which an electric particle describes a five-dimensional geodesic.

However, I was not satisfied with this result but did much work, implying lengthy calculations, on establishing the rigorous equations, which took me most of the summer of 1925, after our return to Denmark. To my surprise they came out to agree exactly with Einstein's gravitational equations and the Maxwell equations in general relativistic form, when the proportionality factor connecting the charge of the particle with the fourth momentum component was appropriately chosen, the sign being such as to make the extra dimension space-like. Taking at that time the geometrical picture of a four-dimensional space literally, this relation, together with the experimental fact that any electric charge is a whole number multiple (positive or negative) of an elementary unit of charge, led me to believe that this space is closed in the direction of the fourth dimension, the circumference being 0.8×10^{-30} cm, far beyond the smallest distances observed. According to this picture, physical quantities ought to be periodic functions of the extra coordinate, measurable quantities being averages taken over this small circumference, the higher overtones corresponding to states of high electric charge. This, I thought, would be the reason why ordinary physical space is only three-dimensional. Moreover, I believed at that time that the periodicity in question was the root of the quantal aspect of nature.

At the end of the summer we went to Sweden to visit my mother before going back to Denmark, where Bohr had obtained a fellowship permitting me to stay in Copenhagen during a leave of absence from Michigan. However, due to my getting seriously ill — an infectious hepatitis — we had to stay in Sweden for a year, arriving in Copenhagen only at the beginning of March 1926. During that half year, when so much happened in physics (Heisenberg's breakthrough in quantum mechanics, Goudsmit's and Uhlenbeck's paper on electron spin, Pauli's matrix theory of the hydrogen atom, to mention the most important ones), I had hardly done any work on reading. But a few weeks before we went to Copenhagen, during a recreation trip, I had written out the summer's work on five-dimensional theory, leaving my quantal speculations for work in Copenhagen. As already mentioned, I started there with the simplest kind of wave equation and tried to work out the stationary states of the harmonic oscillator, when Schrödinger's first wave mechanical paper appeared.

When Pauli came to Copenhagen some weeks later, I showed him my manuscript on five-dimensional theory and after reading it he told me that Kaluza some years before had published a similar idea in a paper I had missed. So I looked it up but — as with de Broglie's thesis, which Bohr had shown me in the summer of 1925 — I read it rather carelessly but quoted both, of course, in the paper I then wrote in a spirit of resignation. In the summer of 1925 I had found Kramers in a state of despondency after the failure of the Bohr-Kramers-Slater statistical interpretation of quantum theory — although, at the same time he had given an important and beautiful contribution to correspondence theory by his dispersion formula. I tried to cheer him up by pointing out in a letter that certainly I considered science a subject of great importance, just as important as the play of children. We had both small children at the time. I thought that now I had to apply this to myself.

In the paper I tried, however, to rescue what I could from the shipwreck, and at the same time to learn as much as possible from Schrödinger and also from de Brogile, whose beautiful group velocity consideration impressed me very much even if by and by I saw that it did not essentially differ from my own way by means of the Hamilton-Jacobi equation. From Schrödinger I learnt in the first place his definition of the non-relativistic expressions for the current-density vector, which it was then easy to generalize to that belonging to the general relativistic wave equation. In this, after Schrödinger's success with the hydrogen atom, I definitely made up my mind to drop the possible non-linear terms, although I was still far from certain that this was more than a linear approximation. Also I derived the energy-momentum components, which in the five-dimensional formalism belonged to the current-density vector. These I published much later, due to the appearance in the meantime of a paper by Schrödinger containing the corresponding non-relativistic expressions.

agree with experimental facts, as is seen from my paper mentioned above (Elektrodynamik und Wellenmechanik vom Standpunkt des Korrespondenzprinzips), the writing of which proceeded very slowly under numerous discussions with Bohr. Still, I hoped, as is seen from its last section, that a further development of five-dimensional theory would give a foundation for this approximate treatment. Hence, after the paper was finally off to the *Zeitschrift für Physik*, I began to look more closely into this matter. First, I convinced myself that Schrödinger's emission theory leads to the Rayleigh formula and not to that of Planck, in agreement with what Bohr and Heisenberg had maintained. Then I tried to see whether anything like the quantal treatment of the many-particle problem might come out of five-dimensional theory. But the result was discouraging.

Then there came a new paper by Dirac containing an application of the quantization procedure on a radiation field considered in the Rayleigh manner as a system of uncoupled oscillators corresponding to its resolution into standing waves. This made me try to use the same procedure on a Schrödinger field coupled electrostatically to itself by means of a scalar potential derived in the usual way from the density. I preferred this approach, starting from a space-time formulation, to that based on a quantum theory translation of general Hamiltonian mechanics of a system of particles. Apart from a disturbing term, where the non-commuting factors had the wrong order, the result of my calculations agreed with Schrödinger's wave equation in configuration space. This was in March 1927, just a year after I became acquainted with the new quantum mechanics. From then on I was convinced that a generalized field theory, whether four- or five-dimensional, ought to be a quantum field theory and not a theory of the classical type.

In the autumn Jordan, who independently had started the same approach came to Copenhagen. He showed me that my trouble had to do with the electrostatic self-energy, which in this particular case was cancelled by an interchange of the non-commuting factors, and we wrote our joint paper. About the same time, I wrote a new paper on five-dimensional theory, where I took the standpoint just mentioned — to the satisfaction of Pauli who had been very angry when reading my correspondence theory paper.

Jordan's and my paper treated only the case of symmetric quantization but, while we wrote it, Jordan told me of his approach to antisymmetric quantization, later treated in detail by him and Wigner. This impressed me immensely. During the past year I had tried to consider the Pauli principle in the wave theory way but had not been able to find a mathematical expression for this all-or-nothing situation, which was now so beautifully realized by Jordan's approach.

Not long after this, Dirac sent Bohr the first draft of his paper on the electron, which came as an amazing surprise. And early in the new year Bohr sent me to Cambridge to learn more about it. This impressed me so much that for a time I abandoned completely my general relativity speculations. And when Pauli, about Easter time, came to Copenhagen we drank a bottle of wine on the death of the fifth dimension — which for a time he and Heisenberg had used in their joint, not yet published work on quantum electrodynamics. But we both had our come-backs. That spring Nishina returned to Copenhagen after a stay in Hamburg and we worked until late in the summer on the Compton effect according to the electron theory of Dirac.

After a short attack of "five-dimensional" in the summer of 1935, I had a more violent one in 1937 — due to the Yukawa theory — on which I gave a paper at a conference in Warsaw in 1938, published in its proceedings. My trouble was, as in my earlier papers on five-dimensional theory, that I tried prematurely to connect the theoretical attempt with the very insufficient knowledge of what is now called elementary particle physics. This time I replaced the periodicity hypothesis — according to which the fields would be represented by Fourier series in the extra coordinate — by means of a two-row matrix representation, corresponding to zero and unit charge only, indicated in a note to the 1927 paper. Moreover, for the first time I used the general relativity form of the Dirac equation as the starting point for a generalized theory. Due to my abandonment of general relativistic speculation after the appearance of Dirac's electron theory, I had paid very little attention to this form, developed in detail by a number of physicists. But more and more I have come to regard this general relativistic equation as the natural starting point for a generalized quantum field theory, also containing non-gravitational interactions, in the first place the electromagnetic ones.

Leaving these rather speculative ideas, I shall say a few words on my position regarding the cosmological attempts of Einstein and his followers. An analysis of Einstein's way to general relativity theory shows that its essential foundation is the principle of equivalence, according to which the ordinary laws of physics, i.e., those of special relativity, must hold in a frame of reference freely falling in a gravitational field which is practically homogeneous within the region and during the time of using this frame. Now, in the first place, such a frame has no relation to the universe at large, being practically realized by a satellite at any place above the earth. Secondly, the principle of equivalence implies that physical quantities get their meaning through measurements made in a frame where gravitation may be neglected. This applies to the mass of any particle or body, having therefore no relation to the totality of mass in the universe, as assumed by Mach.

THE LIBERTY OF DOUBT

Now, the main basis for Einstein's cosmological attempts was this very appealing idea of Mach, and he seemed not to notice its incompatibility with the principle of equivalence. As I have tried to show in some earlier papers and in a paper soon to be published, this takes away the *a priori* arguments for relativistic cosmology, on which Einstein put so much weight. Hence, the interesting *a posteriori* arguments — the so-called fireball radiation, predicted from cosmology by Dicke and the helium content of certain stars, predicted from a cosmological model due to Gamow — means a challenge to those, who like me, try to replace cosmology by a theory on the lines of ordinary physics, regarding the system of galaxies as the first known specimen of a higher-order type of stellar systems.

I should like to finish this talk on unfinished attempts by stressing the great admiration I feel for Einstein's contributions to physics, in the first place to relativity theory and quantum theory, which is certainly not incompatible with the observation that he also shared the universal condition of mankind of making mistakes when trying something new. A study of the history of science — not the history of philosophy — shows that the natural attitude of a scientist is to be inspired by the great predecessors, just as they themselves were by their predecessors, but always taking the liberty of doubting when there are reasons for doubt.

LANDAU—GREAT SCIENTIST AND TEACHER

EARLY TALENT

Professor Lifshitz said:

It is a great honour for me to be speaking here to this audience and it is also a great satisfaction since I can pay honour to the great man who for many years has been the teacher and friend of myself and of many of my Russian colleagues who are present tonight.

Lev Davidovich was born on 22 April 1908 in Baku. His father was a petroleum engineer working in a Baku petroleum firm. His mother was a physician who had done research work in physiology for some time.

He finished at grammar school when he was only 13 years old. He was then already interested in exact sciences and he soon displayed a talent for mathematics. He learnt mathematical analysis by himself and, later on, he used to say that he could hardly remember not being able to differentiate and to integrate.

Lev Davidovich went to the Baku Economic Technical School for one year, as his parents considered him too young to study at a university. From 1922 he simultaneously attended two faculties of the Baku University, the faculty of mathematics and physics and that of chemistry. Eventually he gave up his education in chemistry but his interest in that discipline continued throughout his life.

In 1924 he joined the physics department of Leningrad University. There, at the main centre of Soviet physics at that time, he first met with the stormy development of modern theoretical physics. He went on to study with all his ardent young enthusiasm and he was sometimes so tired that he could not sleep because he kept seeing formulae.

UNBELIEVABLE BEAUTY

Later he said, that at that time he was amazed by the unbelievable beauty of general relativity. He also told about the ecstasies which he experienced over the papers of Heisenberg and Schrödinger giving birth to quantum mechanics. He said that he enjoyed them not only for their scientific beauty but also for their power of human ingenuity, the biggest triumph of which was the ability to understand things which one cannot imagine, like the curvature of space-time or the uncertainty principle.

In 1927 he finished at the university and became a state aspirant of the Leningrad Physico-Technical Institute where he was an "extra state aspirant" from 1926. His first papers were published during these years. In 1926 he published a paper on the theory of line intensities for diatomic molecule spectra and one year later, a paper on a damping problem in quantum mechanics, in which he introduced for the first time, a density matrix for characterizing a state.

However, his enthusiasm for physics and the first success in his scientific life became clouded, due to his oversensitive shyness in contact with other people. He suffered from this and sometimes, as he said later, it drove him to despair. Later, he became more cheerful and felt at ease anywhere and at any time. These changes are due to his characteristic self-discipline and his sense of duty to himself. These qualities, together with his sober and self-critical mind, helped him to become a man with a rare ability for happiness. As he was sober-minded, he was always able to distinguish real things from the idle and unimportant ones and to keep his self-control during difficult moments in his life.

In 1929 he visited foreign countries. He worked in Denmark, England and Switzerland for 1½ years. His stay at the Institute for Theoretical Physics in Copenhagen was the most important. There, together with theoreticians from all over Europe, he attended the famous seminars of Niels Bohr on the fundamental problems of theoretical physics of that time. The scientific atmosphere as well as the great personality of Niels Bohr appear to be decisive for the formation of his outlook on the physical world. He always considered himself as a pupil of Niels Bohr. He visited Copenhagen twice, in 1933 and in 1934. When he was abroad, he published two papers: "On the theory of diamagnetism of an electron gas" and "On a generalization of the uncertainty principle for a relativistic quantum theory" (together with Peierls).

After his return to Leningrad in 1931, Landau worked at the Physico-Technical Institute in Leningrad. In 1932 he moved to Kharkov, where he became head of the theoretical department of the recently established Ukrainian Physico-Technical Institute and, at the same time, a leader of the theoretical physics department of the Kharkov Institute of Engineering and Mechanics. From 1935 he was head of the physics department of Kharkov University. His stay at his university was very fruitful. There he finished for example, "A theory of photo-electrical effects in semiconductors", "A theory of dispersion and absorption of sound", "A study on the behaviour of metals at very low temperatures" and he derived a kinetic equation in the case of Coulomb interaction. Here, he started his teaching activity and he founded his school of theoretical physics.

TEACHING — HIS VOCATION

In theoretical physics, there are many brilliant names of creators in the 20th century. Among them is that of Landau. However, his influence upon the development of theoretical physics consisted not only of his personal contributions but also of excellent teaching. Teaching was his vocation. In this sense, we may compare Landau only with his own teacher — Niels Bohr.

Since his studies at the university, he was interested in the problems of teaching theoretical physics. In Kharkov, he began to work out a programme of "Theoretical Minimum". This was an assembly of the basic knowledge in theoretical physics which was necessary for any experimental or theoretical physicist to be active in fundamental research. In addition, he also taught at the university with such enthusiasm for new methods in teaching physics, that he became head of the physics department of the University of Kharkov. (Later, after the war, he continued to lecture in physics at Moscow University.)

In Kharkov he decided to write a course in theoretical physics and another in general physics. Landau's life-long aim was to write books in physics at different levels. He wrote textbooks for secondary schools as well as specialized lectures for university students. At the time of his car accident, he had almost completed his course in theoretical physics; the first volumes of both courses were in fact finished. He intended to write a course of mathematics for physicists. He thought of it as a course which would give the physicists the necessary knowledge for solving physical problems and therefore this course would have been without mathematical complications and rigour. Unfortunately, he could not realise this programme. Often he emphasized to physicists the importance of being well up in mathematical methods. He thought that a physicist should be able to apply these methods, at least in general cases, so skilfully that his attention would be fully concentrated on the physics of the problem. This required a thorough training, but the status of mathematical education at universities indicated that this training was not always sufficient. We have learnt by experience that the study of mathematics by physicists doing their research, is considered to be too "tedious". That is why Landau required a thorough knowledge of mathematical calculus from everybody who wishes to become "his" pupil.

With this knowledge, the student was allowed to learn seven sections of the "Theoretical Minimum" which included the basic knowledge of all parts of theoretical physics. This knowledge was required before starting any further specialization. Of course, he did not require from anybody such a polyvalent knowledge as he had. In his opinion, theoretical physics was a single science with a common method of work to all its parts. At the beginning of his career, Landau personally examined everybody in the "Theoretical Minimum". Later when the number of candidates had enormously increased, his co-workers also began to examine the candidates. Nevertheless, he always kept for himself the first meeting with young students. Anyone who wished to have a discussion with him could do so, one only had to ring him up and express one's desire.

Of course, not all those who started to study the "Theoretical Minimum" were so bright and so persistent as to pass all examinations. During the period 1934-1961, only 43 students succeeded. The efficiency of this selection can be seen from the following: among those 43 who passed Landau's "minimum", seven were members of the Soviet Academy of Sciences and 23 "Doctor of Science" physicists.

In autumn 1937 Landau moved to Moscow where he became head of the theoretical department of the Institute for Theoretical Problems. There he worked until his death; and it was there that his polyvalent activity reached its peak and he created "The theory of quantum liquids" in a remarkable cooperation with experimental physicists.

INTERNATIONAL RECOGNITION

His merits also gained their due recognition at the Institute for Theoretical Problems. In 1946 he was elected as an active member of the Soviet Academy of Sciences. He was decorated with many honours including two "Order of Lenin" prizes and he received the title of "Hero of Socialist Labour" — an award recognized, not only for his scientific achievements, but also for his contributions to practical solutions of government tasks. He received the Government Prize three times and the Lenin Prize in 1962. There were many honours from other countries as well. In 1951 and 1956 he was elected a member of the Danish and Dutch Academies of Sciences. In 1959 he became a member of the British Physical Society and in 1960, a foreign member of the British Royal Society. In the same year, he also became a member of the National Academy of Sciences of U.S.A. and of the American Academy of Sciences and Arts. Finally, he won the Nobel Prize in 1962 for his "pioneer investigations in the theory of condensed matter, especially of the liquid behaviour".

Landau had a great scientific influence not only on his own students. He was as deeply democratic in his scientific life as in all his life, he was never pompous and had no respect for rank. Everybody could ask him for advice and his answer was always clear and sharp. Only one condition had to be fulfilled — to speak about the actual problem only and not about the academic one which he did not like at all. His intellect was sharply critical and this, together with his deep understanding of the physical meaning of the matter, made discussions with him so attractive and effective.

In discussions, he was ardent and sharp, but not rough; witty, but not caustic. On the door of his room at the university one could read: "L.D. Landau, Warning — he bites!"

When he became older, his temper and manners became softer but his scientific enthusiasm without compromise, remained the same. In fact, beyond his outer sharpness one could always see his impartiality, great humanity and goodness.

Scientific personality and talent brought Landau to the position of the highest scientific referee for his students and co-workers. Undoubtedly, this part of his activity as well as scientific and ethical authority contributed to a considerable extent in establishing the high standard of Soviet theoretical physics.

THE STOCK OF GOLD

The source of his knowledge consisted in maintaining a close and constant scientific contact with a great number of his students and colleagues. The specific feature of the method of his work was that he almost never read scientific articles and books for a period of his life beginning with his stay in Kharkov. Nevertheless, he was always aware of anything new in physics. He found his knowledge in many discussions as well as in seminars which he headed. These seminars took place regularly once a week, for almost thirty years and in the last years they became a meeting of theoretical physicists from all over Moscow. To lecture at one of these seminars was a duty for all his students and co-workers and he himself, selected the material for lectures very carefully and accurately. He was interested and competent in the whole of physics.

For the participants of the seminars it was not always easy to follow the ideas of Landau when he switched from one field of physics to another. He never considered listening to lectures as a mere formality. He was not satisfied until the merit of a given work was quite clear, until everything was proved and all the questions "why not in this way", were answered. Due to careful consideration and criticism, many works turned out to be "pathological" and then he lost interest in them completely. On the other hand, papers which included new ideas and results entered in his so called "stock of gold" and Landau always kept them in mind. In fact, it was sufficient for him to know only the main idea of the work and to derive all the results.

It was usually easier for him to obtain results in his own way than to study the details of the author's method. So he reproduced and thought over the majority of basic results in all the regions of theoretical physics. This was probably just the origin of his phenomenal ability — to answer almost all physical questions received. The scientific style of Landau was against the tendency to convert the simple things to more complicated ones, which unfortunately is now so widespread. He asked just the opposite — to convert complicated things to simpler ones; in other words, to explain in the clearest way, the main laws of nature which rule the phenomena. The ability to do it, i.e., in his own words "to make things more trival," was a question of personal pride.

The general feature of the intellect of Landau was to strive for simplicity and order. This we may see, not only in the way he treated serious problems, but in all matters. For instance, he liked to classify women according to their beauty and theoretical physicists as well. The latter classification was according to their contributions to science and had five classes in the logarithmic scale, hence a physicist of the second class, produced ten times more than a physicist of the third class. (The fifth class contained "pathological" cases.) In this scale Einstein was in the half class, Bohr, Heisenberg, Schrödinger, Dirac and some others were in the first class. He modestly considered himself in the 2½ class and relatively later he took himself to the second class.

He always worked very hard (never at the desk, usually on his couch). His main stimulus was not the striving for fame but the inexhaustible curiosity, the inexhaustible passion for knowledge of the laws of nature in their great and small evidence. "One can never be allowed to suit one's own aims or to seek for one's great discovery, it only misleads," he often repeated.

Besides physics, Landau's fields of interest were very wide. Apart from the exact sciences, he was fond, and has a good knowledge, of history. He was very interested in, and deeply influenced by, all kinds of art with the exception of music and ballet.

Those lucky enough to have been his student or friend over the years knew that Dau, as he was called, did not become older. To be in his company was never boring. The brightness of his personality never waned and his scientific power did not weaken. Thus, it is absurd and more terrible that his brilliant activity was broken off in its prime.